FARMING THE WATERS

FARMING
THE WATERS

Peter R. Limburg

BEAUFORT BOOKS, INC.
New York   *Toronto*

Library of Congress Cataloging in Publication Data

Limburg, Peter R
  Farming the waters.

  Bibliography: p.
  Includes index.
  1. Aquaculture.   I. Title.
SH135.L55   1980   639'.0916   80-23362
ISBN 0-8253-0009-6

Published in the United States by Beaufort Books, Inc., New York. Published simultaneously in Canada by Nelson, Foster and Scott Ltd.

Printed in the U.S.A.   First Edition
10   9   8   7   6   5   4   3   2   1

Design: Ellen LoGiudice

ACKNOWLEDGEMENTS

I would like to thank the following persons for their generous assistance in granting me interviews, supplying me with information, showing me around their aquaculture facilities, or critically reading portions of my manuscript.

Dr. Carl N. Hodges, Environmental Research Laboratory, University of Arizona

Dr. Michael Neushul, University of California, Santa Barbara

Dr. Gary D. Pruder and John Ewart, Marine Studies Center, University of Delaware

R. Léandri, Centre National Pour L'Exploitation Des Océans, Republic of France

Jean-Claude Mourlon, scientific attaché, Embassy of France to the United States

Dr. Ossi Lindquist, University of Kuopio, Kuopio, Finland

Dwight Griswold

Dr. James Hanks, National Marine Fisheries Service Laboratory, Milford, Conn.

Dr. Thomas J. Costello, National Marine Fisheries Service, Southeast Fisheries Center, Miami, Fla.

Dr. Spencer R. Malecha, University of Hawaii at Manoa

Dr. John J. Poluhowich, Institute for Anguilliform Research, University of Bridgeport, Bridgeport, Conn.

Dr. S. Tal, Department of Fisheries, Ministry of Agriculture, Israel

Hal Eastman and Jack Mulhall, Long Island Oyster Farm

Dr. James W. Avault, Jr., and Dr. Larry de la Bretonne, Jr., Louisiana State University

Alvin Casimir, Lummi Indian Tribal Enterprises

Rick Hardy, Martha's Vineyard Shellfish Group

John T. Hughes, Massachusetts State Lobster Hatchery

Dr. E. P. Scrivani, Monterey Abalone Farms

Richard Colantuno, Musky Trout Hatchery, Inc., Bloomfield, N.J.

Mike Sipe, Natural Systems, Inc., Homestead, Fla.

Ms. Rhoda Votaw and colleagues, Marine Program University of New Hampshire

Dr. William J. McNeil, Vern Jackson, and Robert Burkle, Ore-Aqua Foods

Andy Merkowski and Jim Fritch, Otterbine Aerators

Innocencio A. Ronquillo, Bureau of Fisheries and Aquatic Resources, Republic of the Philippines

Dr. Bruce L. Godfriaux, Noel de Blois, Mark Evans, Thomas Passanza, and Harold Swindell, PSE&G

Michael Vernesoni, Quinebaug Valley Hatchery

Dr. Robert A. Busch, Christopher Rangen, and Thorleif Rangen, Rangen, Inc., Buhl, Idaho

Dr. Akella N. Sastry, Graduate School of Oceanography, University of Rhode Island

Steven D. Van Gorder, Organic Gardening and Farming Research Center, Kutztown, Pa.

Ronald Zweig, New Alchemy Institute, Woods Hole, Mass.

Dr. Robert R. Stickney, Department of Wildlife and Fisheries Sciences, Texas A & M University

Roger Oberg, Thousand Springs Trout Farms and Long Island Oyster Farm

Dr. Albert F. Eble, Trenton State College

Dr. Michael Castagna and Dr. John L. Dupuy, Virginia Institute of Marine Science

The author regrets that photographs of certain subjects either did not exist or were not available to a private author. He is grateful for the cooperation of at least some of the many sources he contacted.

*The book is dedicated to
my wife and children*

Contents

FARMING THE WATERS

CHAPTER 1

What Is Aquaculture?

There is a good deal of interest in aquaculture, to judge by the number of articles on that subject that surface periodically in magazines and newspapers. Yet many people are not quite sure of what it is.* Aquaculture is farming the water—the raising of water-dwelling life forms for profit or subsistence, much as livestock is raised. It includes both fresh and saltwater organisms and runs the gamut from the glamorous trout to the humble catfish, carp, and tilapia (a prolific tropical fish, of which you will hear much more later). In addition to fish, it includes mollusks, such as oysters; crustaceans, such as shrimp and crayfish; and even seaweeds, used as food in many Asian countries and also a source of important industrial products.

Although aquaculture will not supply the starving multitudes of an overcrowded world with cheap protein, it has some exciting possibilities. Aquaculture can be extremely productive. A well-managed catfish farm can produce six thousand pounds or more of meat per acre, twenty times the amount of beef that can be raised on an acre. Granted, the fish feed will be much more costly, but there is

---

*Aquaculture should not be confused with hydroponics, or "bathtub farming," the raising of vegetables without soil, in tanks filled with nutrient solution.

also less waste on a catfish than on a steer. And with modern techniques, an acre of seawater can produce as much as 205 tons of mussel meat per year, with a protein and vitamin content higher than that of beef. Furthermore, mussels thrive on nothing more than naturally occurring algae in the seawater.

In addition, though the number of locations suitable for aquaculture are limited, it can alleviate the damage done to natural fisheries by overfishing, pollution, and destruction of such important habitats as coastal wetlands.

Aquaculture has some unexpected branches, such as the raising of goldfish and tropical fish for home aquariums, minnows for bait, frogs for scientific laboratory animals, and brine shrimp for pet fish food. At the moment, the minnow-breeding industry in the United States, largely in the state of Arkansas, is several times larger in dollar volume than the much better known catfish industry, and the ornamental fish industry is even larger. However, this book will concentrate on those animals and plants that are raised for food or for industrial products.

Along with the large-scale commercial aquaculture this book will examine, there is also a good deal of small-scale aquaculture: some is done unscientifically but with a good deal of acquired skill in undeveloped nations. Other modest aquaculture projects are carried on in backyards and even cellars in advanced nations such as the United States, by amateurs who do it as a hobby or to help ease the food budget. An ordinary farm pond, if well managed, can supply more fish than the average family would eat. Small-scale aquaculture will be covered in a separate chapter.

No one is certain just when man began raising fish instead of simply catching them in the wild. Egyptian tomb paintings show fish being kept in pools, though these may have been ornamental rather than food fish. Early in the fifth century B.C. a Chinese scholar referred to fish farming as if it were a long-established and familiar practice, and some historians think it may go back several thousand years earlier. By the beginning of the Christian era the Romans were fattening fish in stone-lined pools for the luxury trade—eels and mullet were two Roman favorites.

During the Middle Ages in Europe, monks kept up the art of

aquaculture, raising several kinds of fishes in ponds to vary the austere monastery diet. The gentry, too, had their private fish ponds: Chaucer mentions in his *Canterbury Tales* that the Franklin (a country gentleman) had "many a breem and luce in stewe." The stewe was a pond, not a bouillabaisse, but this was really not aquaculture, since nothing was done to care for the fish except guard them from poachers. All these efforts produced only small quantities of fish. It was not until modern times that large-scale aquaculture got under way.

There were two main reasons for this. Until recently, the demand for fish was easily met by the natural catch from oceans, lakes, and streams. And it was not until the late 1800's that science began to be applied to aquaculture. The few exceptions will be noted.

However, the situation began to change drastically after the middle of the twentieth century. The world's rapidly growing population demanded more fish, while at the same time the numbers of fish dropped off due to overfishing, water pollution, and the destruction of coastal wetlands by commercial development. These salt marshes are the major breeding ground for important species of saltwater fishes, and every time an acre of wetlands is filled in and paved over to created a shopping mall or building site for vacation homes, or dredged to create a port, it is lost irrecoverably as a nursery for fish, shrimps, crabs, or anything else. Freshwater habitats have been lost an an even greater rate.

Under these circumstances, aquaculture began to look more and more interesting as a supplement to the depleted natural fisheries and—no less important—as a profit-making enterprise. Profit is the key to successful aquaculture. No matter how remarkable are the advances scientists achieve in the lab, they will not be carried out in the outside world unless they pay. And, as aquaculture looked more promising, government fishery bureaus began intensive research into the rearing of fish and other aquatic food organisms. Universities quickly followed suit as funds were made available. Scientists and students carried out innumerable experiments to learn such facts as how many parts per million of dissolved oxygen a given species of fish needed to stay alive and well; the composition of rations on which fish would gain weight rapidly; the upper and lower temperature limits between which each species of fish not only survives but puts on flesh; and other factors that are vital for the successful rearing of fish.

At the same time, long-term genetic studies were undertaken to breed strains of fish that were more resistant to disease, faster growing, and more tolerant of less-than-perfect water quality than their wild relatives. Other researchers investigated feeding behavior, aggression among fish (important when large numbers of fish are penned up together), population dynamics, and factors that trigger reproduction. Similar programs were carried on for mollusks and crustaceans. Indeed, scientists and hatchery managers can now spawn many species of finfish and shellfish at will.

Probably the earliest and most primitive phase of aquaculture was simply capturing the young of the desired species in the wild and providing them with a more or less protected habitat in which to grow to maturity. Thus, the oyster growers of northwestern France simply place strings of tiles in the water at spawning time and wait for oyster larvae to settle on them, exactly as they have been doing for centuries. When the baby oysters reach thumbnail size, they are stripped off the tiles and transplanted to special beds protected from storms and strong currents. There they grow to table size, feeding on the naturally occurring algae in the seawater.

The next step involves providing the fish (or other animals) with food and controlling their environment to some degree. For example, in the Philippine Islands, where milkfish are a popular delicacy, the fry (very young fish) are captured in nets and traps, then carefully moved to specially designed ponds with sluice gates for changing the water and maintaining the proper degree of brackishness. The fish farmers also fertilize the ponds and cultivate a special mixture of algae for the fish to eat—unlike most fishes, milkfish are herbivores. The Filipinos are skilled fish culturists, but the system still has the same basic weakness as all methods that do not breed their own young: in a bad year, there may be very few wild young to catch.

A complete system of aquaculture is self-contained; that is, it must include some way of reproducing the fish or other organisms that are raised rather than depending on an unpredictable Mother Nature to take care of the supply. Spawning your own fish has another ad-

A hand-woven bamboo screen effectively separates fish in a series of ponds on a Philippine fish farm.

vantage besides assuring the supply of young. You can undertake a selective breeding program to breed in the characteristics you want, just as breeders of horses, cattle, and chickens have done for centuries. In fact, one of the goals of modern intensive aquaculture is to produce a truly domesticated fish, as domesticated as any chicken. For ocean "ranching," in which young fish are raised in hatcheries, then released into the sea, the goal is a fish that will survive.

A few species of fish, notably the common carp and the tilapias, breed readily in captivity. For the others, a variety of treatments must be used to trick them into reproducing, or else they must be spawned artifically. Water temperature and hours of light can be regulated so as to make the reproductive systems of the fish go into action, as happens in nature when the bodies of fish respond to natural cues of temperature and day length that are intricately bound up with food supply. Or the fish culturist may shortcut this process and bring his breeding females and males to readiness by injecting them with hormones. In many cases the fish are stripped of eggs and sperm by hand; the eggs and sperm are then mixed in a dish to ensure fertilization. In some species, such as the Pacific salmon, which die after mating, the breeders are "sacrificed"; that is, they are killed before the eggs and sperm are removed.

The first finfishes to be artificially spawned were trout. A German experimenter discovered the method in the 1760's and even published the story in a newspaper. However, the literate public paid no attention, and the technique was forgotten until it was discovered again in 1842 by two French sportfishermen. France led the way with a hatchery in 1854. Other countries followed suit, with both government-owned and private hatcheries turning out baby trout to replenish streams stripped bare by greedy sportsmen. In the first half of the twentieth century, the United States and several other nations created salmon hatcheries to compensate for the many dams that barred the return of salmon to the swift streams where they have their natural breeding grounds.

At the traditional hatchery, the eggs are carefully tended until they hatch, and the fry are fed until they are big enough to release. If to be used for sportfishing, the fish are usually supported at the taxpayers' expense until they are at least minimum legal size for catching, which takes about a year to reach. They are then "harvested" pretty thoroughly by anglers and natural predators. In the case of salmon, the young fish are released at a few months of age and make their way downstream to the sea, where they grow to maturity. (The majority don't make it to maturity, since they are eaten by predators ranging from gulls to killer whales.) Most of the survivors are netted or hooked by commercial fishermen as they return to the spawning grounds.

Although the hatcheries turn out billions of eggs each year and release billions of fish, this is far from a complete system of aquaculture. It is comparable to a chicken farmer turning millions of baby chicks loose in the countryside to fend for themselves and relying on hunters to catch them when they are big enough. But when the hatchery feeds into a culture program, you have a truly self-contained, self-perpetuating aquaculture system. This is true whether each grower breeds his own fish or buys young fish from hatchery specialists to fatten up. The trend in modern aquaculture is toward large-scale, highly mechanized, specialized production, much like the chicken industry. In fact, the chicken industry is one that aquaculturists hold up as a model to work toward.

CHAPTER 2

The Practical Side: Hows, Whats, and Whys of Aquaculture

There is a side to aquaculture that is not revealed in the exciting stories that appear in newspapers and magazines. It is the daily business of keeping the animals alive, healthy, and growing. There is a great deal more to successful aquaculture than just putting fish into a tank or pond. Simply to keep fish and other water-dwelling animals alive, you must pay strict attention to their biological needs. You must also check them constantly to see that they are all right.

Fish are astonishingly sensitive to water temperature. Some species thrive in cold water, others in warm water. You cannot raise both kinds together with good results. Trout, for example, are typical cold-water fish. They can survive down to 32° F. as long as the water does not actually freeze, but they die when the water temperature rises to 75° or 80° F. Their enzymes, the natural catalysts that drive their body chemistry, do not work at these high temperatures. Tilapia, on the other hand, are very much warm-water fish. They can stand water temperatures up to 90° F., but die when the water falls to 50°, a satisfactory temperature for trout. If you try to keep them together in the same tank at some intermediate temperature, say 65°, the trout will eat an inordinate quantity of food (for reasons that will be explained later), while the tilapia will grow poorly because it is too cold for them. It is not enough just to keep the fish alive. They must also put on flesh, and do it at a satisfactory rate so as to reach market size in as short a time as possible. The longer you keep a fish, the more you have to spend on feed, labor, electricity, and other built-in costs of operation. Fish are a crop just as much as hogs, chickens, or corn, and the greater poundage the fish farmer can produce in the shortest period of time, the more profit he will clear. In fact, the professionals think in terms of pounds of fish rather than numbers of fish. They must also think in terms of return on capital investment (facilities, feed, labor, et cetera) and the number of crops per year.

One of the first subjects that comes up when you talk with professional fish farmers is the food conversion ratio. This is the amount of food a fish must eat in order to gain a given amount of weight. Thus, if it takes two pounds of food to make one additional pound of fish, the food conversion ratio is 2 to 1, which professionals consider pretty good. A ratio of 1.5 to 1 is excellent, and not very often attained.

Ideally, a fish would gain a pound of weight for every pound of food consumed, a ratio of 1 to 1. In practice, this cannot happen. Even a sedentary animal like an oyster burns up a certain amount of energy in the mere process of being alive, and a fish, which moves freely about, burns up more to fuel its muscular activity. Food conversion efficiency also depends on how efficient a digestive system the organism has and whether, like a trout, it eats up most of the feed or, like the sloppily feeding eels, it lets half the food fall to the bottom uneaten.

Some other terms that invariably come up when professional aquaculturists are talking are grow-out cycle, dress-out, harvest, and DO. The grow-out cycle is the length of time it takes to bring the fish to market size. It depends on many factors, such as the size of the fingerlings you start with—a five-inch fingerling will reach market size weeks ahead of a three-incher—water temperature, quality of the feed, and the amount of oxygen in the water. In the Unted States, the grow-out period is usually from six to nine months. The shorter the time, the better it is for the fish farmer.

Fish and other water-dwellers have rather narrow optimum temperature ranges for putting on weight. A degree or two can make a significant difference in the progress of a crop of fish. Trout put on flesh most efficiently when the water is between 56° and 59° F. (At temperatures from 60° to 67° they grow faster but are more suscepti-ble to disease, while at lower temperatures they grow slowly and consume more food for each ounce of flesh they gain.) Eggs, on the other hand, often develop better at a slightly cooler temperature, which is why many gigantic fish farms are located where the water is relatively warm, but get their stock from hatcheries in a different part of the country where cooler water favors the breeding of fish.

Dressing out is removing the parts of the fish that are not cus-tomarily eaten, such as entrails, heads, and sometimes the skeleton. Dress-out weight is the proportion of useful meat that is left after the fish is dressed. About 55 percent of a channel catfish, for instance, is edible meat, while 85 percent of a trout is edible. This is about as high as fish culturists have been able to go.

Harvesting is gathering your crop of finfish, shelfish, or whatever when it has reached the desired size. In the case of finfish it can be done by draining the raceway or pond, by netting the fish, or by a combination of these methods.

DO is short for dissolved oxygen, which fish, crustaceans, and mollusks all need to breathe. In nature, dissolved oxygen comes from several sources. A good deal comes from green water plants such as algae and duckweed, which give off oxygen as a result of photosynthesis. However, they can do this only when the sun is shining. Much oxygen passes into the water directly from the at-mosphere. Waterfalls, rapids, and waves bring in additional oxygen by stirring up the water and increasing the surface area that is in contact with the air. In nature, the fish population of a body of water is self-regulating. When there are too many fish for the available oxygen, most of them die off. Since aquaculturists want to avoid fish kills, they often make sure that there is enough oxygen by using aerators. Many fish growers have also found that they can increase their yields greatly by supplying additional oxygen.

Oxygen shortage is most likely to strike on a hot day. The warmer water is, the less dissolved oxygen it can hold; in addition, fish metabolize more rapidly in warm water than in cold water (that is, they are using more energy and their body processes are speeded up), so that they need more oxygen. One of the first signs of oxygen deficiency is that the fish swim up near the surface, where the oxygen concentration is highest. Then they raise their heads above the water to gulp air. When a fish farmer sees this, he knows that he must act quickly or else lose his crop of fish. The recommended treatment is emergency aeration. However, some fish farmers economize by promptly harvesting their fish before they die, processing them, and selling them for whatever they can get. The reason: By the time the fish are gulping air, they may be so severely stressed that they will die within hours anyway.

If oxygen stress (lack of oxygen) doesn't kill the fish outright, it weakens them so that they become vulnerable to the disease organisms that are always on hand in a watery environment. Fish and other water organisms suffer from hundreds of diseases. Some are caused by fungus, some by viruses, some by bacteria, and some by animal parasites. All are quite unpleasant. Normally, fish can resist these diseases pretty well, but when weakened by lack of oxygen, crowding, and other stresses, they succumb rapidly. And disease spreads quickly in a watery environment.

Fish diseases are treated by adding antibiotics to the feed or chemical disinfectants to the water. If carried out in time, such treatments are usually effective. But preventative hygiene is more effective and far, far cheaper.

A strange disorder that sometimes attacks cultured fish is gas-bubble disease, which is not caused by germs at all. It is much like the bends in humans. The bends affects divers and sandhogs, who work under higher than normal atmospheric pressure. Under this high pressure, nitrogen dissolves in the bloodstream. When the diver returns to normal pressure too rapidly, the dissolved nitrogen comes fizzing out of solution like the bubbles in soda water. The bubbles may lodge in the joints, where they cause excruciating pain, or in the blood vessels, where they block the flow of blood and can cause paralysis or death. Fish in a raceway, of course, are not living under abnormally high pressure. But under certain conditions the water can become supersaturated with oxygen, nitrogen, or other gases. The fish absorb the excess gas but they do not pass it out. Instead, it forms bubbles behind their eyes, in their fins, under their skins, and in their blood vessels. A short exposure to high levels of gas supersaturation or a long-term exposure to low levels will eventually kill the fish.

Shipping live fish is almost a science in itself. It is essential that the fish arrive at their destination not only alive but in good condition and unstressed. Most live fish are shipped in specially designed tank trucks, equipped with a battery of devices to monitor and, in the more advanced models, to control oxygen and ammonia levels and water temperature. Oxygen is supplied en route either by mechanical aerators or by the more reliable system of releasing liquid oxygen into the tanks.

Small orders of fish, destined for hobbyists or scientific laboratories, travel in sealed plastic bags with an extra supply of oxygen. (The same technique is used by people who capture tropical fish for aquariums and private collectors.) Live fish being shipped by airplane also travel in plastic bags, protected by a rigid, insulating container.

Fish suffer from motion sickness just as humans do, so they are not fed for at least twenty-four hours before shipping. This avoids the danger of vomiting, which fouls the water and, on a long trip, may cause fish deaths. Tranquilizers are often added to the water to calm the fish and keep them from injuring themselves by swimming about violently in an instinctive attempt to escape this unfamiliar and frightening situation. If possible, the fish are loaded and unloaded by pumping them through a wide tube, which minimizes handling them physically. Otherwise, dip nets are used.

Fish being shipped to a processing plant are good-sized, usually at least half a pound or more. For other purposes they are shipped as fingerlings or fry. With proper prepation, and if none of the life-sustaining equipment breaks down, they can be on the road as long as seven days.

Fertilized fish eggs are also shipped by the millions. The eggs are usually packed in trays and stacked in a plastic bag, with a layer of absorbent material on the bottom and a tray of ice on the top. This keeps the eggs moist and cool, but does not drown them in stale water. It also reduced their need for oxygen chilling them and thus slowing down their metabolic processes. Packed in rigid, insulating containers, fertilized eggs can be shipped all over the world. In fact, there is a substantial trade in fish eggs between the United States, Japan, Russia, and several other nations.

Suppose that you wish to go into the aquaculture business, and that you have learned how to keep your watery livestock alive and healthy. You still have the economic side to contend with. First of all, you must choose a species for which there is a good market. Otherwise you will end up with a lot of wasted seafood, and a big deficit at the bank. People's tastes in food vary widely and irrationally, and there is no accounting for them. It is also extremely difficult to change the public's taste in preferences and persuade them to accept a new kind of food. Thus, catfish is an extremely popular dish in the southern states and the Mississippi Valley. But in the rest of the country very few people will touch a catfish. Why? No one knows. But in consequence you would go broke trying to raise catfish in the Northeast, even if you had a cheap supply of heat. (Catfish grow best at about 85° F.) Could you ship frozen catfish to the South? Yes, but transportation costs would eat up your profits.

Another example: In Kenya, government scientists successfully raised milkfish, which is a prized delicacy in many Oriental countries. They put the milkfish on the market—and the Kenyans refused to buy them.

So you select a species for which there is a good market and a steady demand. Now you must have land on which to place your raceways or grow-out ponds. In many areas, land is scarce and high-priced, particularly near the big urban areas where the bulk of your potential customers live. Also, zoning regulations in densely populated areas may prohibit fish farming on the ground that the wastes pollute water supplies. (Actually, fish wastes are an excellent fertilizer, but up to now it has not been economical to put them into a concentrated form that can be sold to farmers and gardeners.)

Water supply is another key factor. You must have a large and unfailing source of pure water on the order of thousands or even millions of gallons a day—unless you choose to recirculate the water after it has passed through the fish tanks. This has worked well on a small scale in experimenters' labs and in hobbyists' backyard aquaculture systems. However, it is still too costly for use on a commercial scale. It is also prone to mechanical failure, which can result in the loss of the entire crop.

Feed is another major cost. It is perfectly possible to set up a balanced miniature ecosystem that will produce a natural food chain to support the particular species of animal you want to harvest. However, in most cases a natural food chain would not supply enough food for the thousands of pounds of fish you must raise to make a profit and stay in business.

Consider the workings of an aquatic food chain. At the base are the algae and other plants that manufacture their own food from mineral nutrients in the water and the energy of the sun. At the next level are the animals that feed on the algae, which range from microorganisms to fairly sizable fish. (To simplify things, we'll leave out oysters and their kin.) The third level includes all the animals that live on the second-level creatures, and so on. At each level of consumption there is a 90 percent loss, so that it takes one hundred pounds of algae to make ten pounds of second-level animals (larvae of fish and insects, crustaceans, and the like), and one pound of trout. Multiply this several thousand times, and you'll see what a gigantic tank you'd need to support a commercial crop of fish the natural way.

Therefore, commercial fish farmers are compelled to use artificial feeds, which typically contain fish meal (made from cheap and abundant fish like the menhaden and the Peruvian anchovy), vegetable and animal oils, soybeans, grain, and carefully formulated doses of vitamins and minerals. Fish thrive on these feeds, but they are expensive. Researchers have developed feeds that contain no animal protein but as yet they are not competitive.

Labor is another major item. Because of the high cost of labor, most big producers have highly mechanized operations. The fish are fed by machine, and automatic sensors monitor water quality, temperature, and oxygen content. If a lot of hand labor is needed, it may not pay to raise certain species, at least in countries where workers earn a decent wage.

It is apparent that aquaculture is a business, and that it must be done on a large scale to make sufficient profit. In industrially undeveloped countries there are numerous small aquaculturists, who usually raise fish as a sideline. In the United States, hundreds of enthusiasts raise a few pounds of trout or catfish each year in backyards or basements. However, the combined production of both is a small fraction of the world total.

CHAPTER 3

The Glamorous Trout

One of the leading cultured fish in the United States, as in several other nations, is the trout, esteemed for its firm, flavorful flesh. The trout is also a favorite of sportsmen because of the powerful fight it puts up when hooked. Trout are not easy to raise—they are highly sensitive to temperature, oxygen level, and water chemistry—but there is such demand for them, and consumers are willing to pay such high prices for trout, that they are raised in preference to hardier fish. About 95 percent of all the trout that is eaten in the United States is cultured, and about 90 percent of that is raised in a 32-mile section of the Snake River Valley in southern Idaho, where there is a plentiful, year-round supply of water at 58° F., the ideal temperature for trout growth.

Trout, and there are numerous species, belong to the salmon family, or Salmonidae. Three species are raised commercially: rainbow, brook, and brown trout.* All are popular game fish and have been naturalized in many parts of their world far from their native ranges. By far the most important is the rainbow trout, originally native to western North America. The rainbow is more adaptable to hatchery conditions than any other trout species. The brookie is shy and fearful of man. It tends to spook at any disturbance, even being inspected by a man on a catwalk. In nature, this hair-trigger escape response gives the fish an edge over its enemies; in the hatchery, there is no escape, and it leads only to stress and injury. Brown trout, native to Europe, can tolerate warmer water than other species, which is an advantage in many areas. However, their highly aggressive nature can cause problems. Some hatcheries also feature golden trout, a color mutation of the rainbow trout. There is a separate species of golden trout that is native to a few high mountain streams and lakes in California, but is is not raised commercially.

Unlike their cousins the salmons, most trout spend their entire life in fresh water. (Salmon hatch in fresh water but normally migrate to the sea after their first or second year, when they are only a few inches long; they return to fresh water only to spawn.) The salmonids are carnivores and high on the food web. This makes them expensive to feed, since they require rations that are high in protein. Although vegetable-based feeds have been developed, the currently used feeds are still almost half animal protein.

Trout culture began in western Europe about the middle of the nineteenth century. In the United States, hatcheries were established toward the end of the century to stock the streams and lakes depleted by greedy sport fishermen to whom a huge catch was something to brag about as a proof of skill, and by market fishermen to whom a big catch meant a better living. However, trout were not raised for direct consumption in the United States until the 1920's. Many people looked down on cultured trout, considering it to be somehow inferior to the wild product. While there was undoubtedly some snobbery connected with this prejudice, there may also have been some truth, because trout growers had not yet learned that certain algae in the water gave their product a muddy or excessively fish flavor. Today, however, trout are raised under extremely sanitary conditions with strict attention to water quality.

Originally, trout culturists tried to duplicate nature by building large, deep ponds to give the fish plenty of room and using the local

---

*Rainbows and browns belong to the genus *Salmo*; brookies belong to *Salvelinus*, another genus in the Salmonid family.

soil for the sides and bottoms of the ponds. Today, however, long concrete troughs called raceways have almost entirely taken the place of the dirt ponds. Dirt is prone to harbor pathogens and parasites that can wipe out an entire crop of fish in days. The concrete, however, is easy to keep clean and sanitary.

Ponds are also wasteful of space. As the human population grew and spread, and as developed areas occupied more and more of the land surface, it became less and less economical to use large plots of land for trout raising. Realistic aquaculturists learned how to raise more trout in less space. The others gradually went out of business. It was a trout farmer in Idaho who made the revolutionary discovery that a trout, or any other fish, does not need a large volume of water to live in as long as it has enough water flowing past it to supply oxygen and carry off wastes. A trout farm today has its raceways crowded with fish. At feeding time, when the fish concentrate in a few spots, they are packed almost solidly enough to walk on. Average stocking densities range from 2 to 5 pounds of fish per cubic foot of water. At the popular twelve-ounce size, this makes a little more than 2½ to 6½ fish per cubic foot of water. In one experimental project conducted by a New Jersey power company, PSE & G, the culturists have attained densities of 7 pounds of fish per cubic foot of water, and they are shooting for 11. Such densities suggest the crowded conditions under which broiler chickens are raised; they are one way of maximizing profits and increasing yields. Even higher densities have been attained with the durable carp. In Germany, experimenters raised 10 carp to an average weight of 2.16 pounds in a 10½-gallon tank. The photograph of this *tour de force* resembles an aquatic slum; however, the fish obviously thrived.

A trout begins life when a sperm penetrates an egg. In nature, this happens in the clear, cold water of a stream or lake in late autumn or early winter. (Rainbow trout may spawn in early spring, however.) At a hatchery, the "mating" is performed in a shallow pan with the assistance of man. Ripe males and females are stripped by hand of milt (sperm) and eggs, which are mixed together so that fertilization can occur. The fish grower can tell when his fish are ready by picking them up and gently squeezing them. Ripe fish will leak a little milt or eggs from their vents. Unfortunately, each time a fish is handled

Brood-stock trout are dipped from the holding pen to be conditioned for spawning.

increases the chance of disease, due to the breaking of the fish's protective coating of slime. But this is an unavoidable risk.

Since trout growers want to give their fish a head start on the growing season, they try to induce the adults to spawn earlier. This is usually done by manipulating the hours of light and the temperatures to which the brood fish are exposed. Trout are easily fooled by covering their tank with a lightproof cover to create darkness and using electric lights to simulate daylight. However, one to two months are the most that can be gained. If the trout are induced to

Eggs are gently stripped by hand from a ripe female spawner. If properly done, the fish will live to produce eggs for several more years.

spawn too far out of sychronization with their natural cycle, the eggs are not viable. Man cannot always outwit nature.*

The fertilized eggs are placed in an incubator to hatch. There are several types of incubators. Probably the most commonly used is a stack of trays with water entering at the top and passing down through the trays one after another, flowing out at the bottom. Big glass jars through which water circulates are also used, especially in Europe. The most modern type, developed in California, is the upwelling incubator, an open-topped glass cylinder in which water enters at the bottom and flows out over the top into a drain that carries it away. The upwelling water provides excellent circulation and oxygenation. A typical cylinder is twenty-four to thirty-six

---

*It is now technically possible to make trout spawn at any time of the year by using the above-mentioned techniques, plus selective breeding and treatments with hormones. However, this is not generally done.

inches tall and perhaps twelve inches across, and holds about 2 million eggs.

In nature, the eggs develop very slowly at the near-freezing temperatures of the ambient water and hatch in spring, when the natural food supply begins to proliferate. In the hatchery, the eggs take 28 days to hatch at a constant temperature of 52° F., 35 days at 50°, and 50 days at 46°. The last is seldom used, since the aim of the game is to speed up the growing process. These figures are for rainbow trout, which have proven the most adaptable to culture. Figures for other species vary slightly.

The eggs develop faster in warmer water because the warmth speeds up their metabolism, but the temperature limits are very narrow. Even 56° F. is too warm for successful hatching, for most of the eggs die. However, even under ideal conditions, there is a small percentage of egg mortality. A dead egg turns white; while live, healthy eggs are a sort of orange-red; thus, they are easy to distinguish. Dead eggs should be removed from the incubator every day, for they serve as breeding places for fungi that then infect and kill live eggs. An alternative measure is to run a weak solution of formaldehyde or malachite green through the incubator daily to kill fungus spores. This saves many man-hours of tedious and demanding labor searching out and removing dead and infected eggs.

The upwelling incubator saves labor in egg culling, since it carries away dead eggs automatically. Healthy eggs are denser than water and tend to sink to the bottom of the cylinder; dead eggs are lighter than water, and float up and away with the wastes.

After a brief period of "hardening" in water, trout eggs can be handled without endangering them for about forty-eight hours. They then enter a delicate period in which handling may kill them. About halfway through the incubation period, the embryo fish inside the egg develops prominent eye spots. The egg at this stage is called an "eyed egg." Eyed eggs can survive transportation quite well, and millions are sold to hatcheries. The actual hatching takes about a week. Normally the tail is the first part of the baby fish to emerge from the egg. If the head comes out first, the probability is high that the fish is defective and will soon die. For some as yet unknown reason, defective embryos hatch earlier than normal ones.

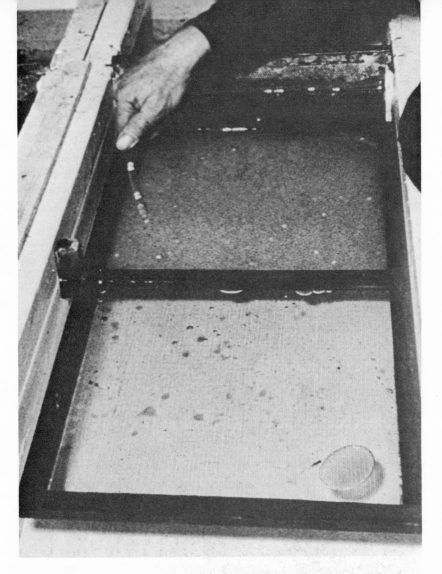

A hatchery worker removes dead eggs from the incubation tray with a suction tube.

Eyed eggs: the eyespots of the developing embryos are clearly visible.

A newly hatched trout is barely recognizable as a fish. It is tiny—only about 9/16 inch long—and most of it is a fat yolk sac. It resembles a tiny football with a head and tail, and it is so lacking in pigment that it is almost invisible. (This may be a natural adaptation to conceal it from predators during this particularly vulnerable stage.) It takes about four thousand of these tiny fish to make a pound. The newly hatched fry, or baby fish, are called *sac fry* because they still have their yolk sacs.

For their first two weeks or so the newly hatched fry spend most of their time lying inert on the bottom of their tank and completing their development. (In nature, they would be hiding under the gravel of the stream bed.) During this period they eat nothing, depending on the food reserves contained in the yolk sac. In fact, they could not eat if they wanted to, for their digestive tubes are closed off. Gradually they grow, acquire better defined fins, and become darker in color. They begin to move around, although they tend to huddle together in one spot. If startled, as by a person dipping a net into the tank, they scurry about like disturbed ants.

By the time the fry are about ¾ inch long, the yolk sacs are abosrbed, and the fry enter the swim-up stage. Their digestive tubes open up, and they begin to eat. They swim actively, searching for food and no longer hugging the bottom. Since their mouths are so small (in nature they would be feeding on the smallest plankton), they require finely granulated feed. They must be fed ten to twenty-four times a day at first; if they are not fed regularly and often, the fry may lose their ability to feed and starve to death. But feeding patterns are soon established, and the trout farmer can reduce the frequency of feeding. By the time the fry are three inches long, they need only six feedings a day. In the final grow-out stages, trout may be fed only twice a day, although many hatcheries feed smaller amounts three to five times a day so that less food will be left uneaten.

As the fish grow, they can handle increasingly larger food pellets, and fish-chow manufacturers have a whole graduated series of sizes that range from a fine powder to ⅜ inch. The composition of the feed also changes as the fish grow. Initially, they require a very high protein and fat content to build flesh rapidly—one successful aqua-

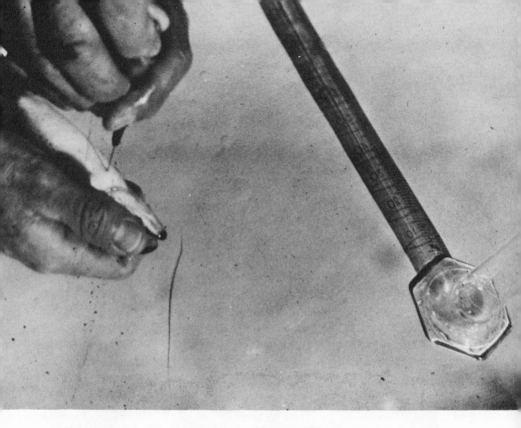

A staff biologist at Thousand Springs takes a blood sample from a young trout as part of a nutritional analysis.

COURTESY OF THOUSAND SPRINGS TROUT FARMS, INC.

culturist told me he uses ground salmon with added vitamins. In later stages the fish require progressively less protein, and the rations contain increasing proportions of carbohydrate, which serves as an energy source so that the fish don't have to burn up protein. Experimenters are working toward a 21-percent-protein diet to lessen costs.

When very small, trout are fed by eye rather than by the textbook chart allowances. By seeing how much remains uneaten, the culturist can tell approximately how much to reduce or increase the feedings. When they reach approximately two inches (and weigh about 1,500 to the pound), the culturist switches to feeding by the chart. Chart allowances are fantastically variable because they are based not only on the body weight of the fish but also on water temperature. The higher the temperature, the greater the per-

Trout are periodically graded for size by making them swim through slatted boxes with apertures of varying widths. Fish of each size class are placed in separate raceways to obtain a uniform crop. Here the graded fish are being dipped up to be weighed.

centage of body weight that is called for. For two-inch fry, for instance, you would feed the equivalent of 3.3 percent of body weight per day at 40° to 45° F., 5.3 percent at 52° to 54° F., and 6 percent at 61° to 63° F. For a four-inch fish, the allowances at the same temperatures are 1.6 percent, 3.2 percent, and 4.0 percent of

36

body weight. It is twice as expensive to feed trout at the upper end of the temperature range as at the lower end. On the other hand, the trout grow twice as fast, providing a quicker return on the grower's investment.

Over 90 percent of all the trout raised in the United States come from a thirty-two-mile stretch of the Snake River Valley in southern Idaho. Here unbelievable qantities of pure spring water gush from the north wall of the Snake River Canyon, at a year-round, unvarying temperature of 58° F. This is ideal for fattening trout, although too warm for spawning them. As a result, the Idaho hatcheries buy their stock as eyed eggs from hatcheries in other areas where the ground-water is cooler.

A visit to the "Magic Valley" of Idaho, as the local chamber of commerce calls it, is a fascinating experience. The Snake River has cut a wide canyon through a bleak, flat, arid plain. As the river wore its way down through the volcanic bedrock, about two hundred feet below the canyon rim it intersected an aquifer (water-bearing layer) of porous volcanic rock, the source of the springs. The aquifer is fed by rain and melting snow from mountains far to the northeast; it is also fed by the famous Lost River, which sinks into the ground, never to reappear. The rock of the aquifer is so porous that oxygen from the atmosphere penetrates it and is absorbed by the water as it percolates its way south. Thus the water arrives at the canyon wall saturated with oxygen, another bonus to the trout farmers.

The Snake continued cutting its way down, and the present canyon floor is perhaps one hundred feet below the aquifer. In certain places one can stand by the road and see cataracts of water gush from the sheer rock wall at rates of many thousands of gallons per minute. One champion spring that was pointed out to me yields 700 cubic feet of water per second, about 314,160 gallons per minute—enough to fill a swimming pool 100 feet long by 50 feet wide and 8½ feet deep throughout. Such huge volumes of water make possible hatcheries that consume anywhere from 4,200 gallons a minute to 140,000 gallons a minute and produce from 600,000 to 5 million pounds of trout a year.

Many of the jetting springs have been captured in huge conduits and tapped for power generation, and wells drilled for irrigation also draw heavily from the aquifer, but the effect of the remaining springs is still extremely impressive.

Between the cliffs and the river sprawl numerous hatcheries, one after the other, like adjacent squares on a Monopoly board. Built at different times, they show technical differences in design that look unimportant to the layman, but have quite an effect on costs of operation and thus on final profits.

For example, hatcheries built in the 1960's have a roadway between each pair of raceways, so that every raceway is accessible by truck or on foot for its whole length. Hatcheries built in the 1970's tried to save money by building raceways with common walls and no space in between. Although there are tracks built across the raceways in several spots, only those immediate areas can be reached conveniently. The rest is very difficult to reach, for cleaning, inspection, or just plain checking on the fish, which should be done every

One of the famous springs of Idaho's "Magic Valley" gushes from the wall of the Snake River Canyon.

COURTESY OF THOUSAND SPRINGS TROUT FARMS, INC.

day. Considering that raceways in Idaho are anywhere from eighty to two hundred feet long, that leaves a lot of inaccessible water. Could the hatchery workers walk along the tops of the walls to check the fish? Perhaps, but the walls are too narrow for good footing. Could they paddle down the raceways in a rubber raft? Yes, but no one does, for it would scare the fish and stress them, and any stress slows growth and lowers the disease resistance of the fish.

Are the hatcheries laid out to gain maximum benefit from the fall of the water? By terracing the raceways so that water can fall from one to the next, free aeration is obtained. However, most of the hatcheries are built on the flat between the canyon wall and the river, and they lack this advantage. Couldn't they use mechanical aerators to make up for this? Technically, yes, but the cost of the electric power needed to run the aerators would bankrupt the hatchery, not to mention that a power failure could wipe out the fish in thirty minutes from lack of oxygen. And so on almost ad infinitum.

The largest hatchery in the state, Box Canyon Hatchery, sprawls over a huge spread on the south side of the river, where land is more plentiful and easier to acquire. It has 200 raceways 20 feet wide and 200 feet long, plus nine more of triple width (60 x 200). It requires 325 c.f.s. (cubic feet per second) of water to maintain an adequate flow in its raceways. The springs, however, are on the far side of the river. Engineers solved this problem by designing an inverted siphon to tap springs on the north wall of the canyon and bring the water under the river. Another economy practiced at Box Canyon and other big hatcheries involves a device to catch the "fines" from the fish feed storage bins. Fines are tiny particles of fish feed, unavoidably crumbled off the pellets. They cannot be used because they are too small for trout to eat and because they clog the fishes' gills. However, collected in bulk quantities, they are sold back to the feed mills.

When it is time to harvest the fish, they are crowded into a pipe by means of a moving barrier, flushed down into a size grader, and then into waiting trucks to be hauled to the processing plants. Everything is done by gravity. Many of these features are used in other hatcheries, too, in order to stay competitive.

Several hatcheries have their own processing plants, which also handle fish from other hatcheries. At Thousand Springs, fish are pumped into concrete holding ponds, from which a suction pipe conveys them to hoppers in the plant. Here the fish are killed quickly by suffocation, the most commonly used method. (Other methods are electric shock and supercooling in very cold brine.) Next, the fish carcasses are fed to a gutting machine, which automatically slits open the fish's belly. A vacuum tube sucks out the viscera and dumps them in a waste bin. A series of automatic knives and brushes remove the gills and whatever viscera escaped the vacuum tube. The wastes are sold as an ingredient in mink food. The gutted carcasses are rapidly chilled, and a moving belt takes them to a weighing and sorting machine.

Some of the fish are packed without further treatment. The rest are hand-boned by a crew of women who work with amazing skill, removing the spines and ribs from inside the fish in such a way that it does not show. The boned fish are sealed in plastic bags, and packed in five-pound master cartons before being sharp-frozen (quick-frozen). As a convenience for the restaurant market, much of the fish is breaded or stuffed with a mixture of corn bread, mushrooms, onions, shrimps, and crabmeat—an example of how far aquaculture can ramify. A good many truckloads of trout are also shipped to market fresh (chilled but not frozen). Altogether, the packing plants of the Magic Valley process about 30 million pounds of fish a year.

At the other end of the scale from the giant Idaho fish farms are the small, individually owned hatcheries that survive mainly by catering to specialty markets. One such is the Musky Trout Hatchery of Bloomsbury, New Jersey.

This hatchery, one of the oldest in the United States, is now owned by Dick Colantuno, a former fish biologist with the New Jersey Fisheries Laboratory, and a few close relatives. With three small installations near the Delaware River, it produces between 80,000 and 100,000 pounds of trout a year. Of this, a large proportion is sold as eyed eggs, fry, and fingerlings to other hatcheries. (Musky's springs yield water at 52° F., which is ideal for hatching but a bit too cold for fast growth.) Dick also sells fingerlings to the EPA and other agencies who use them to monitor water quality in streams and lakes. Dupont, American Cyanamid, and several other chemical producers buy trout for use in bioassays. He sells live two-year-old trout, weighing from twelve to sixteen ounces, to a small number of restaurants within easy trucking range. He sells a good deal of trout, including three- and four-year-olds that measure up to twenty-four inches long, to rod and gun clubs. A few especially choice specimens are bought by photographers for use in advertising or special magazine photographs. A limited number of trout are sold for fishing derbies that various organizations sponsor.

In various parts of the United States there are small "Mom and Pop" enterprises called fishout ponds. Here, for a fee, the public can angle for trout, bass, catfish, or whatever else the proprietor stocks. Fishout ponds are usually only an acre or two in size and are significant mainly in terms of recreation, not food production. However, they are still a part of fish farming, albeit a tiny one.

Trout-rearing practices in other countries do not differ greatly from those used in the United States, although dirt ponds are more used in Europe. In Denmark, where the trout industry is based on a huge supply of saltwater "trash fish," which provide cheap and nutritious feed, only dirt ponds are used. Since suitable spring water is scarce in Denmark, the ponds are laid out along certain of Denmark's many small rivers, and what runs out of Farmer A's ponds flows into Farmer B's downstream. This system is quite vulnerable to disease, although strict sanitation and hygiene keep the likelihood of outbreaks low. But it has happened that all the trout farmers on a particular stream have had their fish stricken by an epidemic, and have had to drain their ponds, destroy the fish, and disinfect the soil of the ponds before starting over.

In Norway, trout are being raised in cages in salt water.* Although trout are freshwater fish, they can be acclimatized to salt water. In fact, the steelhead trout, beloved of sport fishermen, is nothing but a rainbow trout that has gone to sea. The cages keep the fish from straying away and protect them from predators. They also make harvesting the fish a simple matter.

These Norwegian trout are raised in fresh water for their first year. Then water of gradually increasing saltiness is pumped into their

---

*This is also being done on the New England coast and at a few other places in the United States.

tanks for about a month, until the fish are able to tolerate full-strength seawater. At this point they are moved to the sea cages by truck or boat. Here they spend another eighteen months, until they are harvested at a size of 2½ to 3½ pounds. As in Denmark, the trout are fed on fish meal made from "trash fish" that people won't eat, although the commercial fishermen cannot avoid netting them. Ground-up shrimp shells are added to the diet to give the flesh an appetizing pink color—a thrifty use of a waste product from Norway's shrimp canneries.

CHAPTER 4

The Anadromous Salmon

One of the most popular food fishes is the salmon, whose abundance generally makes it a relatively low-priced, staple food item, utilized even in cat food.* At the other end of the scale, fresh poached salmon is a high-priced delicacy found at the finest restaurants, and the best grades of smoked salmon sell for over thirty-five dollars a pound, but that is hardly a part of the mass market.

There are two genera and seven or eight species of salmon. The Atlantic salmon belongs to the genus *Salmo*, along with the rainbow trout and brown trout. It reaches an average size of fifteen pounds, but may surpass eighty. The six or seven species of Pacific salmon belong to the genus *Oncorhynchus*, which is scientists' Greek for "swollen snout." Scientists believe that millions of years ago, ancestral Atlantic salmon migrated across the Arctic Ocean into the Pacific, where their descendants developed the special characteristics that set them apart. Atlantic salmon were once very numerous, but overfishing and destruction of habitat by pollution and dams have sharply reduced their numbers, so that today the great bulk of the salmon catch comes from the Pacific species.

---

*To be accurate, only the waste products are used in pet foods.

PACIFIC SALMON SPECIES

Common Names	Scientific Names	Lifespan	Average Size
Chinook, or king	Oncorhynchus tshawytscha	2 to 7 yrs.	20 lbs.—occasional giants surpass 90 lbs.
Coho, or silver	O. kisutch	2 to 5 yrs.	6 to 12 lbs.; may reach 30 lbs.
Sockeye, or red	O. nerka	3 to 5 yrs.	5 to 7 lbs.; may reach 15 lbs.
Chum, or dog	O. keta	2 to 6 yrs.	13 lbs.; may reach 44 lbs.
Pink, or humpback	O. gorbuscha	2 yrs.	4 to 5 lbs.
Cherry	O. masu	3 to 4 yrs.	5 to 6 lbs.
Cherry*	O. rhodurus	3 to 5 yrs.	5 to 6 lbs.

*Both species of cherry salmon are limited to small regions of Asian waters. They are small fish and of little commercial importance. Many scientists consider them to be two varieties of the same species.

Unlike their cousins the trouts—with one or two exceptions—salmons spend most of their lives in salt water, although they hatch in fresh water and return to it to spawn. Fish with this type of life cycle are called *anadromous*, from the Greek roots *ana* (up) and *dromein* (to run), so named for the way they push their way upstream to spawn, bucking powerful currents and surmounting fierce rapids and waterfalls.

Salmon lay their eggs in cold, clear, fast-flowing streams with gravel beds. With undulating motions of her body and tail, the female scoops a pit in the gravel a few inches to a foot and a half deep and deposits several hundred eggs in it. At the same time the male salmon, who has stationed himself nearby and driven off rival males, presses close to her side and releases a cloud of milt. Although salmon sperm survive in the open for less than a minute, this is long enough to fertilize some 98 percent of the eggs under favorable conditions. The eggs, heavier than water and temporarily sticky, settle to the bottom of the nest pit, and the female covers them again. She repeats this procedure three or four times, moving upstream each time. Gravel that she or other females dig up at each new site is carried down and adds extra covering to the earlier nests. Spawning takes place in summer and in fall, a behavioral adaptation whose function is not known.

The eggs hatch slowly beneath their protective gravel cover, which also permits oxygen-bearing water to circulate down to them. If a flood dumps silt on top of the gravel, cutting off the water circulation, the eggs die of suffocation. Floods may also wash out the nest, scattering the eggs to their doom. About 90 percent of the eggs die before hatching, from these and other causes.

The fry emerge from the gravel on a dark night, which gives them some concealment from predators. Their dark-spotted coloration, mimicking the gravel, also serves to camouflage them. Even so, the odds against the newly hatched fry are severe. Not only floods, but drought and changes in the water temperature can do them in. Predators such as trout, bigger salmon, sculpins, birds, and dragon-fly nymphs take a heavy toll. It is estimated that only 3 to 10 percent of all the eggs laid make it even to fingerling size.

Young pink and chum salmon move downstream to the sea as soon as they are able to fend for themselves. Since these species usually spawn near the river mouths,* the journey is seldom arduous. The others, depending on species and individual variations, remain in fresh water from a few months to four years, slowly developing and approaching readiness for the run downriver. The time of departure is controlled by water temperature and day length, as so much fish behavior is. As they travel, their bodies slowly change to cope with a saltwater life.

A freshwater fish has body fluids with a salt content higher than that of the water it lives in. It wages a constant battle to retain these salts and get rid of the water it absorbs through its gills and the lining of its mouth. A saltwater fish has the opposite problem. Living in water that is considerably more salty than its body fluids, it must retain water and get rid of excess salts. Therefore the kidneys of the young migrating salmon change their structure to do this new job, and their gills develop special cells for excreting salt. In salmon, this

*In the Yukon system, chums spawn as far as 1,500 miles from the sea. Pinks may travel 200 miles upstream.

process is called smoltification, and the newly adapted fish is called a smolt.

The salmon remain in the sea anywhere from one to five years, again depending on their species and individual rates of maturation. They feed on a great variety of shrimplike animals, squid, and fishes, including their own kind. Herring, fatty and high in energy, is a favorite prey. In turn, the salmon are eaten by sharks and other fish, seabirds, seals, and killer whales, and fishermen capture them by the millions. Thus, fewer than 10 percent of all the fry that began the journey to the sea and less than one percent of all the new-laid eggs survive to return to their spawning grounds.

At a certain point in their life cycles, the salmon begin to mature sexually and return from the open sea to their natal streams. For much of their journey they may be guided in part by the Earth's magnetic field, though this has not yet been proven. At any rate, they are able to reach their own rivers from as far as two thousand miles away. Approaching the estuaries, they are guided by the smell and taste of the water. As fry, they are imprinted with the particular chemical signature of their spawning stream, and they recognize this tiny component of the river's outflow. The water of each creek and brook is subtly different from that of all others. So finely tuned is the salmon's chemical sensory apparatus that only one in a hundred or so fails to choose the stream where it was born.

Once the salmon leave the sea, they stop eating and live on their stored-up fat. About one in every ten Atlantic salmon survives the spawning run and returns to the sea, to repeat the cycle as many as four times. The others die of exhaustion and disease after fulfilling their reproductive function.

The Pacific salmons all die after spawning. In preparation for the final act, they undergo strange changes. Males and often females change from the concealing colors they wear in the sea to conspicuous hues with lots of brilliant red.* This may help them to visually identify other salmon ready for mating. The jaws of the males lengthen and hook grotesquely; although they bristle with great, sharp teeth they are useless for eating. The males use their formidable armament only to battle rival males at the spawning sites. Male salmon also develop a hump behind their heads. The size of the hump varies with species. In the case of the pink salmon, it is downright bizarre.

After spawning is completed, the exhausted fish linger near the nest site as long as their strength holds out. Within a week they will either be eaten by the predators that flock to the streams or will succumb to fungus infections such as *Saprolegnia*, which is also one of the banes of fish culturists.

Since the power dams built on the rivers of the Northwest during the 1930's and 1940's drowned many of the spawning streams, state and federal hatcheries were built to compensate for this. Here selected females and males taken from the year's spawning run are killed (they would die soon anyway), and the eggs and milt are surgically removed. The eggs, artificially inseminated, are incubated as with trout, and the fry are released while very small, except coho and chinook, which are released at smolt size. Mortality is high with this method, but it avoids the cost of feeding and caring for the salmon to the fingerling stage.

A more recently developed technique is the artificial spawning channel. This is a man-made waterway, carefully constructed with the ideal size of gravel and water temperature and flow. In the first season, the channel is usually stocked with fertilized eggs. The hatchlings become imprinted on the channel and return to spawn there. The first sizeable spawning channel was built in British Columbia in 1954. Since then, channels have become significant in increasing the numbers of sockeye, pink, and chum salmon.

Whether spawned naturally or in a hatchery, salmon fingerlings must run a gauntlet of dangers to reach the sea, particularly if they must pass a series of dams. The dams have fish ladders, a stairlike series of pools that enable adult spawners to pass the dam going upstream. The little fingerlings, however, usually fail to locate the entrance to the top of the fish ladder. That leaves them two alternatives: either to let themselves be sucked into the intake pipe for the turbines, which kill one out of ten fingerlings passing through, or to

---

*Spawning coloration varies among species and with sex. Generally, the females are much less colorful than the males.

plunge many feet down the spillway. The survivors, resting in the pool below the dam, encounter water that is supercharged with gases, and many succumb to gas-bubble disease.

This is where intensive culture in its various forms comes in. In

Top: Milt (seminal fluid) is milked from the vent of a male coho salmon into a sterile glass flask.

HUGH G. BARTON COURTESY OF THE WEYERHAEUSER COMPANY

Bottom: Ripe eggs are removed surgically from a female salmon that has been selected as broodstock.

HUGH G. BARTON COURTESY OF THE WEYERHAEUSER COMPANY

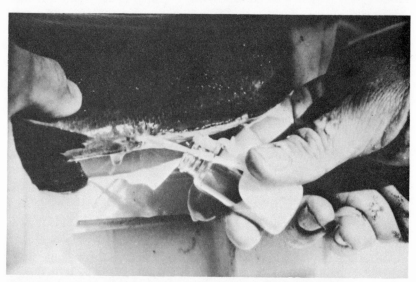

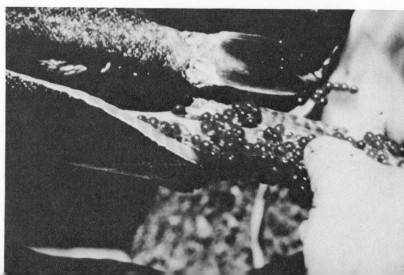

Milt and water are added to eggs to fertilize them. The mixture will be stirred gently with a paddle to ensure maximum exposure.

COURTESY OF THE WEYERHAEUSER COMPANY

Japan, nine-month-old salmon smolts are put in a protected saltwater bay and raised in floating net cages anchored to the bottom. Fed on a high-protein ration fortified with vitamins, these fish in another nine months reach their market size of about 14 inches and 1.1 pounds.

49

A variety of approaches have been tried in the United States. Experiments in raising fry in saltwater lagoons failed because the natural food supply was insufficient and because the oxygen level of the water plunged after algae die-offs. Projects for raising salmon in raceways did all right until intestinal bacterial infections struck in summer. However, Domsea Farms, a subsidiary of Campbell Soup, is successfully raising coho salmon to pan size (about one pound) in net pens, much like the Japanese salmon growers. A number of entrepreneurs in Maine have entered the same business, using coho eggs imported from the West Coast. But government subsidiaries for aquaculture are hard to come by in the Northeast, and Maine's salmon growers must finance their own businesses by outside activities such as lobstering and herring fishing.

One of the most promising innovations is salmon ranching, practiced on the West Coast. Salmon ranching involves raising salmon to smolt size at a hatchery, then turning them loose in the sea to feed themselves, just as cattle are grazed on rangeland. Several firms are engaged in salmon ranching, but the biggest program is that of the Weyerhaeuser Corporation, a giant lumber producer with diversified subsidiaries.

A Weyerhaeuser division called Ore-Aqua Foods operates a hatchery on the Mackenzie River, a fierce-flowing tributary of the Willamette, just outside Springfield, Oregon. Here Ore-Aqua raises 10 million each of chinook, coho, and chum salmon fry yearly, and plans to expand. Each of these species was chosen for a reason. Chinook grows to the largest size of all the salmons and has the finest flesh. The coho is the best adapted to hatchery conditions. The chum salmon is a surprising choice, for its very oily flesh was formerly used for the lowest grade of canned salmon. Early fishermen disdained it, deeming it fit only to cut up for chum to attract better fishes or to feed to the dogs. But chum salmon can be released at a very early age, when only 1 to 1½ inches long and weighing one thousand to the

Aerial view of the Ore-Aqua salmon hatchery at Springfield, Oregon. This hatchery is small compared to the giant trout farms of Idaho's Snake River Valley.

50

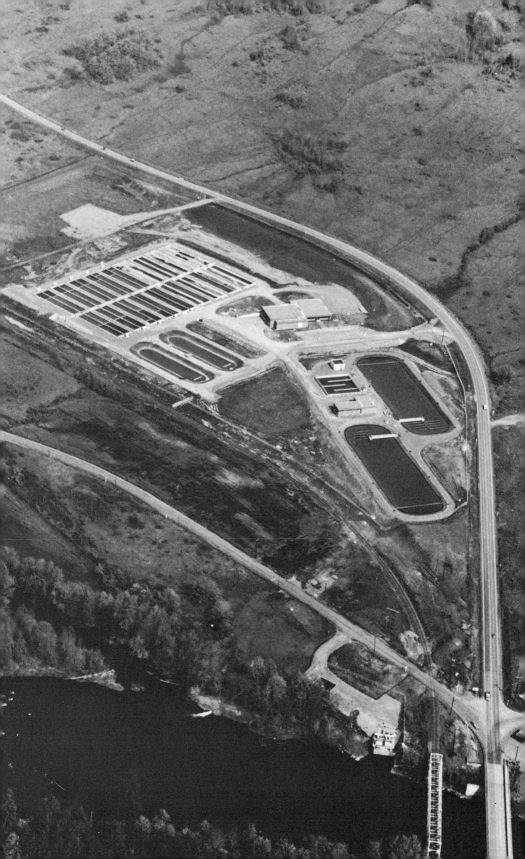

pound (less than one gram or .035 ounce apiece), while chinook and coho are not ready for release until they weigh thirty to the pound. The short period of time they spend at the hatchery cuts the investment in feed and labor to a minimum. They return at an average weight of twelve or thirteen pounds, fattened by nature at no cost to the salmon rancher. Their roe, esteemed as a delicacy in Japan, is a valuable by-product.

At the hatchery, water is taken from the Mackenzie River, chlorinated to kill pathogens, and dechlorinated again so as not to kill the fish; it is then piped into the raceways, mixed with warm effluent from a nearby Weyerhaeuser paperboard mill to raise it to 56° F., an optimal temperature for salmon growth. The fry are fed a ration developed by Oregon State University fisheries scientists, and it gives a food conversion ratio of 2.1 to 1. The combination of nutritious diet and heated water boosts the growth rate of the fry so that in six months they grow as much as they would in eighteen months in the wild. At six months they are ready to become smolts.

As soon as the fish show the outward signs of this metamorphosis by changing color, sample fish are killed and their internal organs examined microscopically. If the findings show that smoltification is complete, the fish are scooped up in nets and checked for condition by hatchery assistants. About one fish in twenty is fed into a tagging machine by hand. This whirring, clicking device inserts a tiny piece of coded wire into the nose of each fish. When the fish is caught as an adult, the tag is recovered with the aid of an electronic scanner. It is decoded under a microscope to identify the year the fish was released and the hatchery from which it came. (Once in the sea, the fish become public property, free for the taking by anyone who can catch them.)

After tagging, the young salmon are loaded into tank trucks and taken to the release ponds on the coast. Ore-Aqua has two release facilities: One is at Newport, on the mouth of the Yaquina River, and the other is at Coos Bay, on a lonely sandspit behind a pulp mill. Both are about one hundred miles from Springfield, over steep, twisting mountain roads. Here on the coast the most ingenious part of the salmon-ranching cycle takes place: The fish are imprinted with the chemical signature of the release ponds.

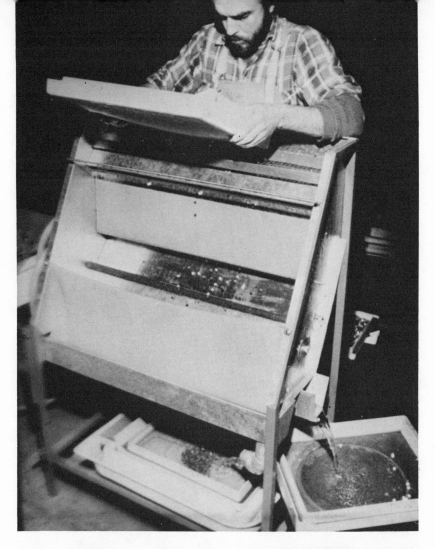

A technician decants eyed eggs into a machine that separates viable eggs from dead eggs. Good eggs bounce into collecting troughs beneath machine; dead eggs (white) do not bounce and go into disposal container beside machine.

COURTESY OF THE WEYERHAEUSER COMPANY

The developing salmon embryo looks like this at eight weeks.
COURTESY OF
THE WEYERHAEUSER COMPANY

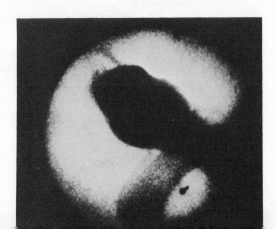

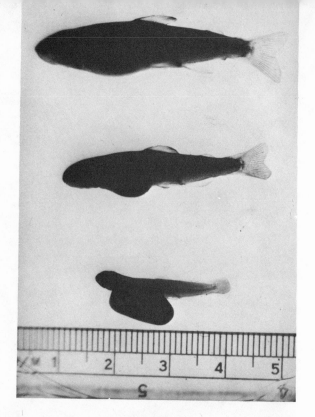

The yolk sac is gradually absorbed as the fry develop.

HUGH G. BARTON COURTESY OF THE WEYERHAEUSER COMPANY

Salmon fry in pre-smolt stage. Note that fins are transparent, a feature that may aid in camouflaging the fry against gravelly river beds.

COURTESY OF THE WEYERHOUSER COMPANY

Top: Juvenile salmon are vaccinated by stirring them in a vat of water to which the vaccine has been added. Each batch is stirred for approximately 30 seconds.

Bottom: At the Ore-Aqua hatchery, randomly selected smolts are tagged before release by this machine, which inserts a tiny bit of coded wire in each smolt's nose. The procedure does not harm the fish. An alternate method of tagging is to clip the second dorsal fin.

Opposite page: Smolts are flushed out of the end of the raceway into a tank truck for transportation to the holding pond where they will be acclimated before release.

Top: At Ore-Aqua's release facility at Newport, Oregon, smolts are pumped into the holding ponds to become imprinted with the water's chemical signature.

Bottom: Returning salmon swim up the several gently inclined ramps of this fish ladder to reach the holding ponds on whose water they were imprinted in Ore-Aqua's ocean ranching program.

Conservative practice would call for putting the smolts into diluted seawater and gradually increasing its strength, but Ore-Aqua's scientists have found this unnecessary as long as the smoltification is complete. A few smolts die of the shock of transplantation into full-strength seawater; the rest recover in a few hours. The smolts are held in the release ponds for ten to fourteen days, at the end of which they are indelibly imprinted on the water of these ponds. At the peak of a high tide, the big outlet pipes of the ponds are opened, and the salmon are sucked down into the bay and swept out to sea by the falling tide. Most of the salmon are imprinted and released in late spring. The slow growers are kept at the hatchery over the summer, to be imprinted and released in the fall.

After reaching maturity in the sea, the salmon return to the release ponds, drawn by the same powerful instinct that guides wild salmon to their natural spawning streams. The coho return after 18 months at sea, the chinook and chum after 2½ to 3½ years. There is also a certain percentage of so-called jacks—undersize adults that have matured ahead of schedule. The function of jacks in nature is not clear, but they turn up in every spawning run.

The Ore-Aqua salmon swim up a gently inclined fish ladder to their ponds and then pass through a sorting channel. Here the brood stock is selected by a staff member seated above the channel, who makes a quick visual inspection of each fish as it swims by. A flip of an

A salmon swims over size-grading bars to enter the channel leading to the holding ponds. Salmon are counted electronically at this point.

COURTESY OF THE WEYERHAEUSER COMPANY

A hatchery technician displays a plump coho salmon that has returned to Ore-Aqua's Newport release facility after a year and a half at sea. Hooked snout shows this fish is a male.

electrically operated gate separates them: the biggest and best-looking fish are directed into a holding pond for breeders; the others go to another pond for processing. As of this writing, Ore-Aqua is giving priority to building up its own selected brood stock. For the first few years, they had to purchase eggs from state hatcheries, since they lacked a sufficient pool of mature adults. In 1978, chum salon eggs were virtually impossible to obtain; so it was necessary to purchase 10 million from a Russian hatchery on Sakhalin Island. However, Ore-Aqua plans to reach self-sufficiency in a few years.

The brood fish are held for six to eight weeks, until they are fully ripe. They are then killed and the eggs fertilized on the site. After a brief period for water-hardening, the eggs are trucked back to the hatchery at Springfield to begin the cycle anew. After the eggs and milt are removed, some of the breeders are of good enough quality to be sold for human consumption. The rest are ground up and processed into fish feed for the hatchery. The salmon that are destined for sale are harvested mechanically and killed by suffocation—this process gives the most all-around satisfactory results in flesh quality. They are then sorted by size and species, iced down, and rushed to nearby processing plants.

Survival of the fish up to the time of their release is better than 70 percent. Of those that return from the sea, commercial and sport fishermen take about 80 percent. However, the Ore-Aqua management estimates that if they recover as little as one percent of all the smolts they release, they will break even, and a 3 percent recovery will mean a comfortable profit.

So far the chief hindrance to salmon ranching has come from commercial fishermen. The reasons for this are unclear. An unreasoning hostility to large corporations appears to play a part, although Weyerhaeuser has gone to considerable lengths to act responsibly. And once the salmon are in the sea, they become public property, free for the taking by anyone who can catch them. Some of the opposition may also be motivated by a greedy and shortsighted fear than an increase in the supply of salmon will bring down the market price. This writer is of the opinion that increasing the supply of this valuable food fish is a highly desirable goal and if it takes a giant corporation to bring it about, so be it. Everyone should benefit in the end.

This has already proved true in Russia and Japan, the foremost practitioners of salmon ranching. Between them, these two nations have already released over *two billion* young salmon. In northern Japan, salmon ranching is established as a major industry.

Ranching the Atlantic salmon may one day be a reality, too. United States federal and state hatcheries along the Connecticut River and elsewhere in New England are breeding small numbers of Atlantic salmon in an attempt to reestablish the runs that once filled the rivers. Although mortality has been high among the released salmon, a few have successfully made the return trip, an encouraging indication. A decade ago this would have been a visionary scheme, but environmental cleanup programs have now improved water quality in many rivers so the point where salmon can once more survive.

CHAPTER 5

Catfish—Unaesthetic
But Delicious

Ugly to behold but good to eat are the catfishes, a large group of fish that are found, in one form or another, from the cooler parts of the temperate zones to the tropics, and on every continent but Antarctica. Some members of the order are sea-dwelling, but the vast majority are freshwater fish. The catfishes range in size from a few inches to twelve feet, and have a great variety of shapes and colorations. However, all of them have fleshy, whiskerlike feelers, called barbels, around their mouths, and sharp spines in their dorsal and pectoral fins (that is, the fins on their backs and those just behind their gills). Most species have poison glands connected to these spines; some are extremely venomous.

Most catfishes live in slow-flowing streams or lakes and ponds. They are predators and scavengers as the occasion arises. Although basically carnivorous, they are not choosy about what they eat, and have even been known to strike at a bare hook. Catfish forage along the bottom for food, aided by their barbels, which contain taste buds.

The most common and widely distributed catfish in the United States is the bullhead, which can reach a size of eighteen inches and three pounds in weight, although they are usually less than half that size in the wild. Bullheads are extremely hardy and can survive in water where other catfishes would die of lack of oxygen (they have an air bladder that acts as a kind of emergency lung, allowing them to breathe air directly). However, they are seldom raised commercially. They are susceptible to disease, are slow growers, and have a poorer food conversion ratio than other American catfishes. Also, their ugly looks render them unappealing to many customers.

By far the leading cultured catfish in the United States is the channel catfish, native to the Mississippi drainage system. The channel cat prefers clear, flowing water but adapts well to ponds. Channel cats are handsome fish, slender in form and light bluish-gray in color. They reach a maximum size of four feet and thirty pounds. Channel cats are good culture fish because they stand crowding well.

Blue catfish, with the same native range as the channel cat, grow considerably larger in the wild, reaching 150 pounds. This is no advantage in aquaculture, however, since the usual market size is one pound. It is axiomatic among fish farmers that the smaller you can harvest your fish, the better off you are, for each new day means another chance for something to go wrong.

Blue cats are also cultured on a fairly large scale. Their flesh is considered slightly better than that of a channel catfish, and they have a better dress-out ratio: 60 to 62 percent of live weight compared to 56 to 58 percent for a channel cat. Blue catfish have the added advantage that they can be trained to feed at the surface, which enables the culturist to see how much of their food they are eating. However, they grow more slowly and don't convert their feed to flesh as efficiently. Also they do not grow as uniformly as channel cats—a greater number are bigger or smaller than the average weight. The fish farmer, however, needs a product of uniform size.

Other catfish native to the United States are the flathead catfish, a native of the Mississippi basin, and the white catfish, native to streams that drain into the Atlantic. The flathead, an aggressive and cannibalistic creature, is seldom cultured. In the wild it grows to a weight of one hundred pounds. The white catfish are not much cultured at present: drawbacks include slow growth and a big head that cuts dressout weight. However, they can stand crowding, low

Former Rodale fish biologist Andy Merkowski and unidentified helper display a gigantic channel catfish, probably a spawner. Normal market size for cultured catfish is much smaller.

COURTESY OF RODALE RESOURCES, INC., AND BARBERO, INC.

oxygen, high temperatures, and turbid water better than channel cats; so it is likely that they will become an important cultured fish in years to come.

Catfish farming in the United States began in the 1950's when a few rice growers in Arkansas and other Southern states decided to grow catfish as a sideline in their flooded rice paddies. Together with the catfish they raised buffalo, a large fish of the sucker family, which is popular in the Mississippi basin area. Although this scheme did not work out, other farmers with plenty of land and water began raising catfish alone. By the mid-sixties, catfish farming was making real headway, and by 1969 the catfish harvest reached approximately 66 million pounds.

It was a story that is becoming familiar in aquaculture. Not only was there a guaranteed supply of fish, but consumers found that the cultured fish was of more uniform, and usually better, quality than the wild fish. In other words, with cultured fish they could be almost 100 percent sure of getting a good meal.

Since catfish must have warm water for fast growth (optimum temperature, about 85° F.). the industry is concentrated in the southern states, where there is a long warm season. Arkansas, Louisiana, and Mississippi lead in catfish production, although there are catfish farms in at least eighteen states.

There is even one lonely outpost in the Snake River Valley of Idaho, where a fish culturist named Leo Ray uses water from a hot spring on his land to raise catfish in raceways. Mr. Ray's raceways march down a high hillside in series like a gigantic stairway, each series twenty-four feet long and ten feet wide. In these raceways Ray

This 12-inch channel cat weighs about ½ pound. Average market size for cultured catfish in the United States is two pounds.

COURTESY OF RODALE RESOURCES, INC.

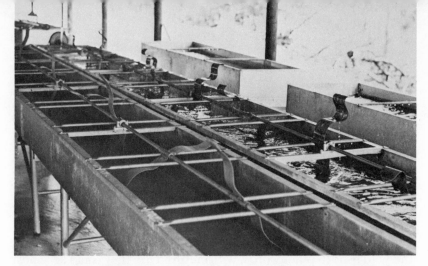

Top: Incubation troughs at Mississippi State catfish hatchery. Motor-driven paddles gently stir water and eggs, duplicating the care given by the male catfish in nature.

MISSISSIPPI COOPERATIVE EXTENSION SERVICE

Bottom: Catfish farming is highly mechanized because of its large scale. Here a tractor-drawn, powered hopper spews feed pellets into a catfish pond. Fish rise to grab the pellets as they splash down at right.

MISSISSIPPI COOPERATIVE EXTENSION SERVICE

Opposite page: A catfish harvest in Missippi's Delta region. Workers crowd fish into a corner of the the pond with a seine; then a crane dips them up and transfers them into a truck.

MISSISSIPPI COOPERATIVE EXTENSION SERVICE

has achieved amazing densities of as much as ten thousand pounds—five tons—of catfish per section. This works out to nearly thirteen pounds of fish per cubic foot. However, he now works with a more manageable density of 6,000 pounds of fish per section. Since he harvests his catfish at a two-pound average size, each section holds three thousand good-sized fish. Seen from above, the raceways are a squirming mass of fish with very little elbow room—or perhaps fin room would be a more appropriate expression for fish.

Experiments in catfish raising have been made in other northern states, using hot effluent from power plants to heat the raceways. In theory, this year-round supply of heat could shorten the growing season considerably, but in practice the results have not been promising as yet.

The great bulk of catfish are raised in dirt ponds that range in size from an acre or two to twelve acres, although some growers have ponds as big as forty acres. The smaller ponds are easier to manage; but one big pond costs less to construct than several small ponds covering the same area. The ponds are kept filled with three to six feet of water.

The fish are stocked in spring as fingerlings, and in six to twelve more months they are big enough to harvest. Catfish are harvested with seines. Usually the pond is drained almost completely beforehand to concentrate the fish in the harvest basin, a deeper spot constructed near the outlet pipe. Sometimes the fish are lured to the harvest basin by feeding only there and nowhere else in the pond. This avoids the loss of stragglers that would otherwise flap about on the drained bottom and die unharvested. (It's just not economical to send a crew of men sloshing through the mud looking for lost fish.)

Unlike the trout industry, which sells most of its harvest frozen and packed, many catfish are sold alive. Many catfish farmers keep a supply of fish on hand in holding ponds, ready for instant loading and delivery to buyers. This means an extra capital outlay for the catfish farmer, but market conditions make it necessary.

A container of live, newly harvested catfish is lowered into a holding tank at a processing plant.

MISSISSIPPI COOPERATIVE EXTENSION SERVICE

Another difference between catfish and trout farming is that many catfish farmers spawn their own fry instead of buying them from specialized producers. The simplest way of spawning catfish is to place a number of brood fish in special breeding ponds, much smaller than the grow-out ponds, and setting out receptacles for them to mate in. Milk cans, the big kind dairy farmers use, make the best receptacles, but nail kegs, earthenware crocks, and similar objects are also used. The males claim receptacles as their territories. The females pair off with them, and in the privacy of their own container male and female consummate their mating. It is a protracted process: the female extrudes batch after batch of sticky eggs from her vent, and the male fertilizes each batch when it emerges. The final result is a spongy mass of stuck-together eggs about the size of a man's fist.

There are certain difficulties with this procedure. Male catfish fight viciously at breeding time, and may seriously injure or even kill one another. Males may also attack females, which is hardly conducive to a successful mating. Once the eggs are fertilized, the male chases the female away and mounts guard over the eggs, fanning water over them with his fins. Sometimes he may stir the egg mass up with his body. The male stays on duty until the fry are able to swim about freely, three to five days after they hatch, which is five to ten days after they are laid.

The receptacles should be checked regularly to see if they contain eggs or fry. One way—not the easiest—is to shoo the male out, lift the receptacle up, and pour out most of the water. This must be done gently, so as not to spill out eggs or fry. An easier technique is to insert a length of rubber hose into the receptacle and feel around. If you encounter a spongy, inert mass, you have found eggs. If you feel a wriggling mass, the eggs have hatched. Sometimes the male bites the hose, considering it an attacker, and he has to be extricated from the receptacle somehow. It is best not to do this by hand, because an angry catfish can deliver a vicious bite.

Before the eggs hatch, the fish farmer must be very careful to disturb the males as little as possible. Catfish are very touchy animals, and even a loud noise such as a car door slamming near the pond can cause some males to eat their eggs in a kind of defense

70

response gone wrong. To avoid this, some growers incubate the eggs in troughs.

Another method is to place individual breeding pairs in pens of wire mesh, each with its own spawning receptacle. The wire must extend about six inches down into the bottom of the pond and a foot or two above the water to keep the fish from burrowing their way to freedom or jumping over the top of the pen. The biggest risk with this method is that the male can easily corner the female and kill her if she is not ready for spawning. The advantages are that the fish farmer can select the particular fish he wants to mate instead of leavng it up to chance, and the spawn gets some protection from predators.

The most efficient spawning technique is to inject the selected females with hormones, which brings them to readiness within twenty-four hours. (Males apparently need no such stimulation.) The breeding pair is placed in a tank, and the eggs are collected on a tar-paper mat for incubation.

The fry are usually started out in rearing ponds, where they are kept until they reach four-inch fingerling size. Then they go into the big grow-out ponds. As fry, they are vulnerable to predators such as frogs and fish-eating insects. They must also compete for food with tadpoles and a kind of freshwater crustacean called fairy shrimp. Another troublesome crustacean is the tadpole shrimp, which stirs up the mud and clouds the water. As fingerlings, young catfish are preyed upon by herons, gulls, kingfishers, and other birds. Most of these birds are protected species, and the fish farmer is not permitted to shoot them. He can, however, attempt to frighten them away by such means as firing a carbide cannon, which produces a loud blast, or by broadcasting the birds' recorded distress calls at very high volume. Unfortunately, the birds soon learn that the noise is not followed by any real danger.

Trout farmers sometimes build wire cages over their raceways to keep out troublesome birds, but the smaller birds are expert at finding gaps to sneak in through. In one case, the hatchery manager left just a one-foot space between the top of the cage and the surface of the water. To his dismay, the birds quickly learned to soar along that narrow space and scoop up a meal.

In Europe, catfish are raised mainly in Hungary and Yugoslavia. The species raised is the native European *wels*, which is found naturally in the Danube River basin and east into Russia. The wels is the giant of the catfish family, reaching a length of twelve feet. Fast growers, they weigh two pounds at the end of their second year. They are usually sold as four- to six-pounders at the tender age of three.

Catfish raising in southeast Asia goes back many centuries. Traditionally, the Asian fish growers captured wild fry in fine-meshed nets and raised them in small ponds, often in combination with other species of fish. The catfish were and still are fed on kitchen scraps, bananas, cooked rice, bran, and soft green plants. The fish are usually harvested at two to four pounds, which they take two years to reach on this largely vegetarian diet. If fish scraps are added to the regimen, they reach harvest size in under a year.

An improvement, invented in Cambodia, is to raise fish in floating cages of bamboo. Sometimes the cages are built into a raft, on which the fish culturist lives. In cage culture, it is standard practice to mix a high proportion of chopped trash fish in with the rest of the feed.

Work on artificial propagation of native Asian catfish began in Thailand in the 1960's. However, most Asian catfish growers still rely on capturing wild fry.

A Thai specialty is the culture of two species of clariid catfishes, a widespread family with gills that have evolved to breath air as well as water. With this ability, they can live for hours out of the water, and they can survive indefinitely in water low in oxygen or even in wet mud. Clariids have been known to come out of the water at night and prowl the shore for food.

One of the clariids raised in Thailand, and so-called "walking catfish," was imported to Florida as an aquarium fish. Some escaped from a dealer's holding ponds, and their progeny rapidly overran a large part of southern Florida. Due to their extreme aggressiveness—in an aquarium tank they intimidate even the dreaded piranha—they quickly kill off most of the native fish in ponds and streams. Nothing seems to stop the walking catfish. If their ponds are poisoned, they simply travel overland to a safer home, partly using the stout spines of their pectoral fins as legs, and partly humping along the ground on their sides, using nose and tail for support. Wildlife Service experts fear that the walking catfish may spread through all of Florida and into neighboring states, but so far it has not.

However, the walking catfish is an excellent food fish, and one day may become commercially important in the United States.

CHAPTER 6

Carp and Tilapia: Two Utilitarian Warmwater Fish

Carp, disdained by American consumers and cursed by American sportfishermen, are highly esteemed in Europe and Asia. They are probably the fish cultured longest by man. Chinese records of carp culture go back to 475 B.C. (a treatise on how to spawn carp), and some authorities believe the Chinese were raising carp as long ago as 2000 B.C. The Greeks and Romans of classical times may have raised carp in ponds, though there are no records of this, and medieval Europeans probably began raising carp on a small scale in the twelfth century.

The carps belong to the family Cyprinidae (from the Latin name for the carp), which they share with the numerous minnow clan. It is a large family, with something like 1,500 species in all. One of the best-known members of the family is the goldfish, domesticated many centuries ago by Chinese and Japanese breeders and refined by them into an endless variety of coloration and shapes ranging from graceful to grotesque. A number of carp cousins that most Americans have never heard of are popular food fish in Europe and are often raised together with carp: the tench, roach, rudd, and ide. The bream is another European cyprinid, not to be confused with the breams of the southern United States, which are actually sunfish.

The giant of the cyprinid family is the mahseer of India, which reaches eight feet in length and often weighs over one hundred pounds. At the other end of the scale are the tiny, colorful barbs and danios of southeast Asia, which are popular with aquarium owners.

The common carp is a rather squat-bodied fish with big scales and two pairs of fleshy barbels on its upper lip, one long and one short. It likes quiet ponds and sluggish streams. Its original range was from eastern Europe to China and eastern Siberia. Centuries ago it was taken to western Europe, where it quickly made itself at home. Carp were brought into the United States in the 1870's and 1880's and gained a foothold over wide areas of the country. They soon gained a bad reputation from their habit of rooting in the mud for food. Sportfishermen claimed this made the water too muddy for some of the popular American game fishes to survive in. The carp was also accused of eating the eggs and fry of native fishes. One reason for the carp's unpopularity is almost certainly the great number of small, sharp bones that turn eating a carp into an obstacle course, and the fact that carp often take on a muddy flavor from certain algae on which they feed. (The muddy flavor, however, can be removed by keeping the carp in clean, running water for a few days.)

Modern biologists point out that carp probably did not do nearly as much damage to native fishes as they have been blamed for. Overfishing and pollution by man reduced the numbers of native fishes, and the prolific, hardy carp filled the gap, being able to survive in water that is muddy, polluted, and low in oxygen. Although carp undoubtedly eat some eggs and fry of other fishes, their main foods are insect larvae, algae, water plants, worms, zooplankton, and decaying plant matter.

In Europe, carp are considered one of the most challenging game fishes. Greedy as they are, they are extremely wary, and anglers take the most extraordinary precautions to approach them. The fisherman annoints his hands with oil of anise before baiting the hook, to destroy the telltale human odor; or he may wear gloves instead. The fisherman's face is often daubed with mud or hidden behind a beekeeper's veil. He wears shoes with thick, soft soles that deaden his footsteps, and when he sights a school of carp he may even sneak up on them on all fours. Whether such elaborate precautions are really necessary is difficult to say; at any rate, they are part of the ritual of the sport. Once hooked, the powerful carp can easily break light fishing tackle, and landing them is a challenge in itself.

Although common carp can survive the glacial winters of Minnesota and Siberia, for purposes of aquaculture they are considered warm-water fish because they grow slowly in water under 64° F. Where summers are hot, they can be raised economically, although they reach market size much faster if the water is warm the year around.

In nature, the common carp can reach a length of 2½ to 3 feet and weigh 20 to 30 pounds; occasional giants of 80 pounds have been caught in nets. But market size for cultured carp ranges from 1 to 4 pounds, except in Indonesia, where people relish tiny carp of 2½ to 4 ounces.

The common carp spawns freely in captivity, and European culturists developed improved varieties with chunkier bodies and finer flesh. Besides an improved scaly carp, there is the mirror carp, which has a single row of scales along each side of its back fin and a few large scales scattered at random over its otherwise naked body. The mirror carp is also called the Israeli carp because it is the principal fish raised in Israel and because Israeli scientists have done a great deal of work on improving the breed. There is also the line

Nepalese carp growers use these cone-shaped earthenware jars for incubators in place of the costlier glass jars used in Europe.

WFP/FOA PHOTO BY E. WOYNAROVICH

Aquaculture trainees in Indonesia select prime specimens of common carp for spawning.

FAO PHOTO BY JACK LING

carp, which has a single line of scales down each side, and the leather carp, which is almost entirely scaleless. European consumers prefer the scaleless varieties, probably because they are easier for the housewife to prepare. However, scaly carp and mirror carp grow faster and have greater disease resistance.

Common carp are raised in Europe under a great variety of conditions, from low-yielding ponds, where they forage for themselves on the natural food web, to intensive culture with feeding, which yields up to 1,300 pounds per acre. In parts of Central Europe, they share their ponds with ducks, whose excrement helps to nourish an abundant growth of algae and other phytoplankton, which in turn feeds other organisms that carp eat. A couple of cities in West Germany raise carp in ponds of treated sewage water, which also supports a rich algae growth. If carp are reared in cages and fed, as is done in several countries, the yield is over 3,500 pounds per acre. The Indonesians place their carp cages in polluted streams, gaining a high but unhygienic yield of over 400,000 pounds per acre. The highest yields of all have been achieved by Japanese culturists with closed, recirculating systems that turn out the equivalent of 3.5 million pounds per acre. However actual yields are much smaller, because of the limited size of the tanks.

In Israel, the only water available for aquaculture is too salty (three parts per million of salt) to use for irrigation. However, the carp thrive in this semi-brackish liquid, and Israeli scientists are working toward developing more salt-tolerant strains. Tilapia and mullet are often raised with the carp, increasing fish output without significantly raising feed costs (in Israel, carp are fed on fish chow).

It is the Chinese, however, who are preeminent in carp culture. In addition to the common carp, they raise several species native to eastern Asia: grass carp, silver carp, black carp, bighead carp, and mud carp. Since these fishes occupy different ecological niches, and even spend most of their time at different levels in the water, the Chinese took advantage of this behavior to establish the world's oldest known polycultural system.

In a traditional Chinese carp pond, grass carp inhabit the upper levels of the water and feed on almost any kind of plant matter, from water weeds to grass clippings and garden trash. In addition, the

An Israeli carp (left) shares the stage with a channel catfish at one of Rodale's small-scale experimental culture facilities.

COURTESY OF RODALE RESOURCES, INC.

grass carp's abundant and incompletely digested feces serve as food for other carps in the polyculture mix. Mud carp and common carp forage diligently along the bottom, eating benthic fauna (bottom-dwelling animals) and detritus, including grass carp feces. Another bottom-feeder is the black carp, which eats snails and other mollusks and crustaceans, crushing their shells with powerful teeth in its gullet. In the midwater levels swim silver carp, which live on planktonic algae that they filter from the water with their very fine gill rakers, and bighead carp, which eat zooplankton.

Top: The grass carp, one of the Chinese carps, is raised in the Orient as a food fish. It is used in some areas of the U.S. to control water weeds.

U.S. FISH AND WILDLIFE SERVICE

FISH FARMING EXPERIMENTAL STATION STUTTGART, ARK.

Bottom: From top to bottom: grass carp, silver carp, bighead carp, and Israeli carp.

PHOTO BY DR. HOMER BUCK

U.S. FISH AND WILDLIFE SERVICE

FISH FARMING EXPERIMENTAL STATION STUTTGART, ARK.

Nepalese fish farmers harvest carp with a seine. Ducks are also raised on this pond in an adaptation of a Chinese polycultural system. The duck excrement fertilizes the water and increases the natural plankton supply for the carp.

FAO PHOTO BY PEYTON JOHNSON

The actual mixture of species varies with climatic region and other factors. The mud carp, for instance, dies when the water temperature drops below 54° F. for any length of time; so it is not raised in northern China. Black carp are not raised on Taiwan because very few Taiwanese will eat them. Silver carp cannot be raised where the water is very muddy, because the suspended soil particles clog their fine gill rakers. In colder parts of China, some fish farms use Chinese bream to fill the role of bottom-feeding omnivore. In coastal areas, milkfish and mullet may be pressed into service as vegetarian plankton eaters.

Chinese carp are traditionally put into the pond as fingerlings and kept for a grow-out period of three years. Since Chinese fish culture is part of a tightly integrated farm system, the ponds are fertilized to increase natural food production. Almost the only supplementary food is the vegetable matter fed to the grass carp. Pig and poultry manure, and human excrement, the basic fertilizer of the Orient, are used. In some cases, outhouses are built over the pond to get the fertilizer there with the least effort. The excrement of the grass carp also serves as manure. The sludge at the bottom of the pond is dug out periodically and used as manure on the vegetables, thus completing the cycle. Much the same system is used throughout southeast Asia.

Unlike common carp, which will figuratively spawn on a wet handkerchief, Chinese carp do not spawn in captivity. Until recently, Chinese aquaculturists depended on wild-caught fry. Since each haul of wild fry usually contains a mixture of species, the tiny fish are separated by pouring them into a tall sorting basket in the water. Within a short time the fry sort themselves out into layers: silver carp at the top, bighead next, grass carp followed by black carp in the midlevels, and mud and common carp at the bottom. The fry are carefully dipped out, a layer at a time, with fine-meshed nets.

Artificial spawning of Chinese carps began in the 1930's with silver and grass carp. The fish were stripped by hand of eggs and milt. Progress was held back by the Japanese invasion and by World War II, but by the 1950's China could supply millions of fry of these species to fish farmers. Even so, the supply could not approach the demand. However, in the 1950's Chinese fishery scientists began to use a technique developed in Brazil in 1934 called hypophysation. It involves injecting an extract of carp pituitary gland, or hypophysis, into mature fish to jolt their gonads into action. With this technique, mass production was made possible, and by the 1960's it was well under way. Other hormone treatments include chorionic gonado-trophin, extracted from the urine of pregnant women, and luteiniz-ing release hormone (LSH), which plays a part in the human men-strual cycle. A synthetic analog of LSH has been manufactured in Chinese laboratories.

With these achievements in spawning technology and their age-old intensive polycultural systems, the Chinese are believed to be the world's largest producers of cultured freshwater fish. Carp will probably play an increasing role in freshwater aquaculture as a cheap source of high-quality protein. Grass carp and silver carp are also being used experimentally to control water weeds and algae, and black carp are being used in parts of Asia to control snails that compete with other carp for food and oxygen.

Common carp prowl the bottom of their tank at the U.S. Fish and Wildlife Service Farming Experimental Station at Stuttgart, Arkansas, an important center of research on warmwater fish.

U.S. FISH AND WILDLIFE SERVICE
FISH FARMING EXPERIMENTAL STATION STUTTGART, ARK.

Tilapia are another workhorse of aquaculture. Like carp, they are unglamorous but tasty, prolific, and hardy. Largely vegetarian, they are cheap to fed, and they have been supplying protein to mankind for many centuries. The ancient Egyptians were catching tilapia, and perhaps raising them in ponds, as long ago as 2500 B.C. The fish that the New Testament tells us the Apostles caught in the Sea of Galilee were almost certainly tilapia, as were those in the story of Jesus feeding the multitudes.

Tilapia—there are about fifty species—belong to the Cichlid family, which are mostly small, brightly colored tropical fishes of a highly aggressive nature. Tilapias, however, are usually silvery-gray or black in color and fairly peaceable in temperament. Native to Africa and the Near East, they have been spread around the world in modern times. Some species are favorites with aquarium fanciers because of their interesting habit of brooding their eggs and young in the female's mouth.

In the wild, tilapias may reach twenty pounds, although one- and two-pounders are much more common. When raised in ponds, they are apt to be much smaller. The reason is that tilapias begin to breed at about two to three months of age, and they are so prolific that they soon outbreed their food supply. The result is a pond full of dwarfed, stunted little fish, all head and no body, weighing only a few ounces but still ready and willing to reproduce themselves. Sexual activity itself slows their growth rate: the energy that goes into producing eggs and milt is not available for growth. In addition, the females do not eat at all while they are brooding eggs and larvae in their mouths, a period that may last as long as three weeks. Strict attention is necessary to keep the population thinned out and permit the fish to grow to marketable size.

Some of the many species of tilapias are omnivorous; some eat zooplankton and phytoplankton. Some specialize in filamentous algae, while others eat macrophytic vegetation, which means plants that are big enough to see without a microscope. Although tilapias die below 45° F., they can survive temperatures of over 112°F. The ideal growing temperature for several commonly cultured species is 89°F. They thrive in water that is high in organic pollutants and low in dissolved oxygen. About twelve species are currently cultured in various parts of the world.

Tilapia, native to Africa and the Near East, are now cultured on a major scale in southeastern Asia and hold great promise for Latin America and the United States.

FAO PHOTO

Tilapia were cultured experimentally in Kenya in the 1920's, the idea being to provide an easy-to-raise fish crop for African peasants. Tilapia are now cultured in Zaire and several other African nations, both for subsistence and for market. However, Africans tend to regard fish as wild creatures that ought to take care of themselves. After all, in nature fish grow without man's care, while gardens and millet fields obviously need to be looked after. Therefore they often neglect their fish ponds, ending up with undersized fish that they consider unfit to eat.

Tilapia first appeared in large numbers outside their native range in 1939, when they surfaced in Java. It is not known precisely when or how they reached that sensuous tropical island, but aquarium owners are suspected. Once released or escaped, the tilapia spread rapidly throughout the island, infesting not only ponds and streams but even ditches and stagnant, rain-filled quarry pits. So prevalent were they that they quickly became known as Java tilapia, a name they keep to this day despite the fact that they are native to the east coast of Africa and that their scientific name, *Tilapia* mossambica*, refers to the former Portuguese colony of Mozambique, on the African mainland.

At first the tilapia were not welcomed by the Indonesian aquaculturists, who accused them of all sorts of misdemeanors, including eating up the special algae mixture that the fish farmers cultivated to feed the highly valued milkfish. However, it was impossible to get rid of the tilapia. Then in 1941 war reached Indonesia. The Japanese army invaded it and stayed there until the end of World War II. Under the brutal and tyrannical Japanese occuaption, the laborious traditional systems of fish culture could no longer be carried on, and tilapia came to the fore because they required so little attention. The Japanese, always interested in a good fish crop, promoted tilapia culture on the other Indonesian islands and also took them to Malaysia. From there the finny African emigrants spread throughout southeastern Asia.

Today tilapia are a major crop in southeast Asia and are raised as far north as Taiwan, where they have even surpassed carp in importance. They are often raised in polycultures with other species of fish, such as carp and mullet. Incidentally, some species of tilapia are able to do quite well in brackish water and even in seawater, which extends their cultural range considerably. Sometimes predatory fish are raised with the tilapias to control their numbers. They are harvested with the tilapias and sold as a by-product. However, this natural kind of population control is a tricky business. Too many predators eat up the tilapia crop; too few may not be able to gobble up the excess of small, unprofitable tilapia. Size ratio of the predator

---

*A few years ago taxonomists decided to reclassify the tilapias, placing all the mouthbreeders in a new genus, *Sarotherodon*, and those that build nests on the bottom in the genus *Tilapia*. Almost all the cultured species fall into *Sarotherodon*, but fish farmers have resisted the change of nomenclature.

and prey is also a critical factor. Predatory fish rarely attack a victim more than half their own size; so the aquaculturist must be sure to add his predators before the tilapia grow too large to be easy prey. A variant system is to use small tropical fish such as the Jack Dempsey, which eat the tilapia fry.

A somewhat surer method of population control is to raise mono-sex cultures, all-male or all-female. The fingerlings are sexed and separated before they reach breeding age—their sexual apertures are different enough to be distinguished. However, the process is slow and laborious, and even experienced fish sexers seldom achieve 100 percent accuracy. Alas, all it takes is one member of the wrong sex to undo all the precautions.

More promising are the experiments that have been done in crossing two species that yield all-male hybrid offspring. There are three combinations known to produce all-male fry, but the most practical is a male *Tilapia hornorum* with a female *Tilapia mossambica*.

In practice, "all-male" stocks often contain a few females. However, experimenters have found that adding predators to eat up the fry takes care of this problem. The same technique has worked well with hand-sexed cultures.

The newest approach to monosex cultures is to treat normal fry with minuscule doses of testosterone and other six hormones. Mixed into their feed for the first five days that they swim free from the mother, the hormones convert almost all of the young females into males. Results of 98 to 100 percent male stocks have been achieved. It will be some time, however, before fry producers are able to supply these hybrids on a large scale.

A simpler method for controlling the numbers of tilapia is to raise the fish in floating cages. When a female lays her eggs, they drop through the mesh floor of the cage to the bottom of the pond or raceway without being fertilized by a male—a nearly foolproof method of contraception.

Tilapia are grown as a warm-season crop in Israel, and they are being considered by fish farmers in our own Gulf States. Fish farmers in Texas, in particular, are beginning to raise tilapia in ponds enriched with manure from animal feed lots. In six months they reach fourteen ounces, a respectable market size. Since disposal of the huge amounts of manure that accumulate at feed lots is a major problem, this is an economical solution that accomplishes two goals at once. The adaptable tilapia can also be raised in sewage-treatment ponds, although this raises certain hygienic questions.

Tilapia also show promise for culture in the heated effluent from power plants and factories. In addition, they are very well suited for raising by backyard hobbyists.

The fact that tilapia thrive on a mainly vegetable diet makes them particularly suitable for the overpopulated, impoverished regions of the world, where protein is too scarce to feed to fish. They gain weight well on grass clippings, vegetable tops, and similar garden trash; and they also relish broken rice, peanut flour, rotten fruit, and the sweepings from flour mills. In a pilot test conducted by Mike Sipe, a Florida tilapia breeeder, tilapia raised in the waste lagoon of the Tropicana Orange Juice processing plant at Bradenton gained 7 percent a day without supplemental feeding. Had the entire lagoon been utilized, it would have yielded better than 100,000 pounds of live fish per acre. Sipe believes that, with proper design, instrumentation, and monitoring, it may be possible to produce over 1 million pounds of tilapia per acre at waste-water treatment plants.

One species of tilapia, *T. melanopleura*, may one day be used to control water weeds. *Tilapia mossambica* and *nilotica* are already used to control mosquitoes by eating the filamentous algae in which many species of mosquitoes love to breed. Tilapia that are too small for human consumption can be ground up for animal feed or fish feed, or used as bait fish. In fact, Japanese tuna fishermen are using *Tilapia zillii*, a species that adapts readily to salt water, as live bait in place of anchovies.

The Florida tilapia breeder mentioned above, Mike Sipe, who is also a master geneticist, has developed true-breeding hybrid strains of tilapia in attractive colors. So far he has produced golden, red, white, and orange fish, both solid-colored and mixed. Marketing tests indicate that these bright colors are popular with shoppers who are put off by the normal gray or black hue of the tilapia. Mr. Sipe has also developed a "dumpy" hybrid that contains a much higher proportion of edible meat than the normal shape. Other goals that geneticists may be working on right now could include a super-fast growth rate, nonaggressiveness, earlier maturity (for producers of fry and fingerlings), and tolerance of cold water. It has even been hinted that a fish with soft, cartilaginous bones that don't interfere with eating is in the works. But the experimenters are not yet talking about these developments to the public.

CHAPTER 7

Marine and Brackish-Water Fishes

Not nearly as much as been done with culturing the fishes of the open sea and the inshore waters as has been done with their freshwater relatives. However, one or two species have been raised by man for centuries, and in recent years more have been added.

The leading cultured marine fish is the milkfish, a native of warm waters in the Pacific and Indian Oceans. Looking much like an overgrown herring, although they are not related, milkfish may grow as long as four feet and weigh over forty pounds. They are an almost ideal fish for culturing: they tolerate a very wide range of salinities; they are disease-resistant and do not seem to mind crowding; they can tolerate low oxygen levels in an emergency; they grow rapidly and have flesh of very high quality; and they feed very near the bottom of the food web, on algae, zooplankton, and small animals such as worms. Cannibalism is no problem, because of their mainly herbivorous diet.

Milkfish culture probably began on the island of Java, where it dates back at least to A.D. 1400. It is also very important in the Philippines and Taiwan, and milkfish culture on a commercial scale is getting under way in Hawaii.

Since milkfish do not spawn in captivity, and artificial spawning

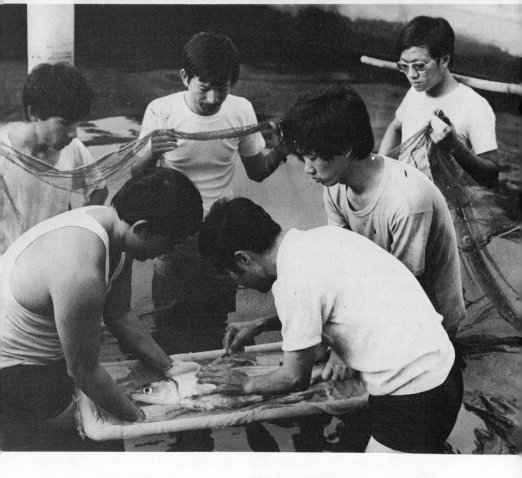

Taiwanese aquaculturists inject a milkfish with hormones to induce spawning.

was not achieved until 1979, culturists have had to rely on wild-caught fry. Fortunately, the milkfish is very prolific. Each female releases about 5 million eggs at a time, and in some areas milkfish spawn twice a year. The tiny eggs float in the warm water close to shore and hatch in about twenty four hours. The larvae, helped by tieds and currents, make their way to tidal estuaries and mangrove swamps, where they take cover behind sandbars, rocks, fallen trees, and other obstacles. Here they congregate, and fry catchers capture them with dip nets. Where there is no natural cover, the fry catchers sometimes build arificial breakwaters. A less laborious method makes use of a long rope called a *blabar* by Indonesians. Made of palm fiber, the blabar has garlands of leaves or grass woven into it all

91

along its length. The fishermen float it out on top of the water and the fry, seeking refuge beneath the garlands, are easily netted. In some areas, the fishermen stake out funnel-shaped baskets of bamboo splits, which trap the fry as the current carries them by.

Once caught, the fry are popped into watertight containers and sold to dealers who often come right down to the beach. Milkfish culture is a highly compartmentalized business, and the fry may pass through half a dozen middlemen before they reach the actual grower.

Milkfish ponds, whether nursery ponds for fry or grow-out ponds for fingerlings, are usually built on the shore, in a mangrove swamp or on a tidal estuary. Most of them do not connect directly with the sea, but are supplied via tidal creeks, ditches, or canals. Preferred practice is to build the pond so that its bottom is slightly below the high-tide level. Sluice gates let water in and out. The water in the ponds is brackish and the salt content changes with every rain.

The pond must be carefully prepared before the fish are put in. Although details vary from region to region, the general procedure is to let the water out at low tide. The mud bottom is then dug up and turned over like a garden plot. Before it dries, it is leveled with wooden rakes. The bottom is given a very gentle slope down toward the outlet, and all humps and hollows are smoothed out.

The bottom is then allowed to dry until the soil cracks, which kills off most of the resident predators. Sometimes the pond has to be flooded and dried several times to get rid of eels and other predators that burrow deep into the mud. The pond is dried again, and tobacco waste or tea-seed cake is spread over the bottom. A few inches of water are let in, and the resulting toxic brew gets rid of the predators that the preceding treatment did not kill off. The botanical poison sources soon break down and provide organic fertilizer for the pond.

Finally, the water is allowed to evaporate once more, and new water is let in, just enough to cover the bottom one to two inches deep. A bamboo screen across the intake strains out most unwanted organisms, although some larvae always slip through. Within a few days a lush mat of blue-green algae, which Filipino culturists call *lab-lab*, begins to form. Although classified as blue-green, the algae are actually brown, yellow or green in color. In addition to numerous species of algae, the lab-lab contains bacteria, diatoms, protozoans, tiny crustaceans, larvae, and worms, which adds up to a very nutritious ration for milkfish of all ages.

As the lab-lab grows, a few more inches of water are added, and the fish are put in as soon as the lab-lab has made sufficient growth. They are stocked quite densely to keep their growth rate uniform. Unlike most fish, young milkfish can be kept in an underfed, stunted state for as long as a year without ill effects. Once released to a less crowded pond and fed adequately, they resume growth at a very rapid rate. This makes it possible for dealers to keep a large stock of fry or fingerlings on hand at all times.

As the young milkfish grow, they are often given supplementary rations of rice bran or starch. In the Philippines, some growers feed their milkfish on a red seaweed called *Graciliaria*, which increases their growth rate. Certain growers have switched to raising *Gracilaria* as the basic milkfish food in place of lab-lab.

Taiwan has the highest yields per acre of milkfish, sometimes surpassing 2,600 pounds per acre. Since winters in Taiwan are too cool for milkfish (they die below 48° F.), those that are carried through the winter are kept in special ponds protected by windbreaks or covered by plastic greenhouses. This protection is not needed in the Philippines or Indonesia.

Taiwanese growers raise monocultures of milkfish. In the Philippines and Indonesia, however, the culturists almost invariably end up with an unplanned polyculture of every kind of sea-dweller that sneaks in through the sluice gates. Fortunately, most of these intruders can also be sold, and 20 to 30 percent of a milkfish farmer's gross in these countries comes from prawns, crabs, mullet and other fish, and tiny crustaceans that are made into a paste that is used as a flavoring.

Mullet have been esteemed as food fish since ancient times. They are mainly found in tropical and semitropical waters, but a few species range far into the temperate zones. Hardy mullet are found in Long Island Sound and off the coast of England. Mullet are found in the Atlantic, Indian, and Pacific Oceans and are also common in the Mediterranean Sea.

Like milkfish, mullet are raised in brackish ponds. Although they are saltwater fish, they can adapt to fresh water. Mullet fry feed on plankton. Adult mullet swallow mud from the bottom and strain out everything edible. Most species have a gizzardlike stomach to grind up their food. Mullet also graze on diatoms growing on rocks and seaweeds. In ponds, they feed on decaying plants.

Mullet are raised in Italy, Israel, and other Mediterranean lands as well as in Southeast Asia and Taiwan. Very seldom are they raised in a monoculture because of uncertain supplies of fry. Although researchers in Israel and Taiwan have learned how to spawn mullet artificially, most of the hatch die in the larval stage. Mullet are most commonly raised together with tilapia or carp. However, once artificial spawning and selective breeding become commercially practicable, mullet may well become a saltwater counterpart to trout.

Yellowtail, a good-sized member of the jack family, are cultured in Japan. Yellowtails are closely related to the tasty pompano. In the wild, yellowtails average about two feet long and weigh about ten pounds.

Yellowtails are not the most likely candidates for aquaculture. They are wide-ranging, fast-swimming fish of the open sea, and they have a voracious appetite for other fish. In fact, they were not cultured on a large scale until 1960. Since yellowtail have not yet been bred in captivity, the fry are captured off the coast of the big southern Japanese island of Kyushu in spring. The Japanese government sets strict limits on the number of fry that can be taken each year, which helps to prevent depletion of the wild stock.

The fry are sold to specialists who raise them to fingerling size in floating cages of nylon netting. Before being stocked in the cages, the fry are graded into large, medium, and small sizes. Only fish in the same size class are placed together; otherwise the large yellowtail fry would quickly eat the small ones. The fry are fed on minced shrimp or white-fleshed fish (a costly diet) until they are from four to six weeks old and about four inches long. They are then transferred to very large floating cages, 42 to 120 square yards in area and 10 to 20 feet deep. Cages have all the advantages previously mentioned of keeping the fish in one place, protecting them from predators, and making harvesting a snap. When crowded with fish, they also increase the food conversion efficiency of the fish by preventing them from moving about and burning up calories.

Although yellowtail is a high-cost, luxury product, it seems likely that other nations will follow the Japanese example and culture their local species of yellowtail.

Little is known about the life history or dietary requirements of the pompano and its larger cousin the permit, both native to the warm waters of Florida and the Caribbean, with an extreme range from Massachusetts to Brazil. There are also two Pacific species. However, researchers from the United States National Marine Fisheries Service, the University of Miami, and other institutions including two giant food corporations have been attempting to raise pompano for years.

Although scientists have been able to spawn pompano by the use of hormones, the eggs usually fail to hatch. Wild-caught fry have a distressing tendency to die en masse at various stages of growth due to factors that are apparently related to their diet. However, the hope of eventual high profits together with the probability of the depletion of wild stocks keeps the work going.

Japanese fisheries scientists have been investigating the possibility of rearing the giant bluefin tuna in floating pens. One problem is that bluefin die when the water turns cold; another is how to supply these big predators with enough of the right kinds of food. An alternative proposal is to screen off the entrances to coral atoll lagoons in the Pacific and use them as tuna ranches. However, every version of this plan has so far involved enriching the natural food web of the lagoons by adding fertilizer. It is feared that intensive fertilization might throw the whole ecosystem of the lagoon out of balance and kill off the coral and the rich and varied biological communities associated with it. So far, however, the magnitude of the task has deterred any experiments in tuna ranching in the lagoons.

One of the most publicized experiments in saltwater aquaculture is the British flatfish project. Flatfish culture actually began in the 1890's, when countries on both sides of the Atlantic set up hatcheries to restock depleted fishing grounds. However, these restocking programs had no discernible effect on yields, and they were discontinued in most countries after a few years. Norway, however, still restocks its plaice fisheries with hatchery fry.

Research on intensive flatfish culture began in Britain during World War II, under the impetus of the Nazi blockade. The fish

Young flounder (a flatfish) in a University of New Hampshire/Sea Grant aquaculture project. the fish have been chemically branded for identification.

COURTESY OF THE UNIVERSITY OF NEW HAMPSHIRE—

MARINE PROGRAM, DURHAM, NEW HAMPSHIRE

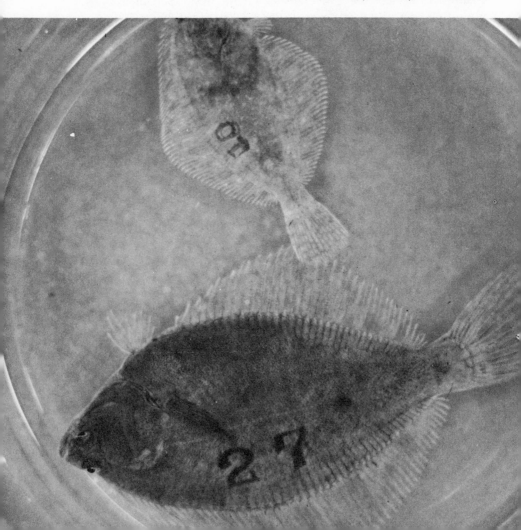

chosen were plaice and sole. Spawning and hatching were easy, but for years 90 percent of the larvae died at a very early stage. Finally, in 1962, a researcher found that the cause of this mortality was a bacterial infection that could be treated with antibiotics added to the water of the tanks. This allowed 80 to 90 percent of the fry to survive to the stage of metamorphosis, when the normal-looking, upright-swimming baby flatfish undergoes the changes that fit it for a life spent lying flat on the bottom. One eye migrates halfway around the skull, the skull itself twists, and in many species the mouth twists around in a 90-degree angle so that the jaws operate from side to side rather than up and down.

Flatfish hatchery operations are now concentrated on the Isle of Man; the hatchery there ships out fry to grow-out ponds at several locations. The brood stock spawns naturally from February to May. As soon as the mating act is completed, a hatchery staffer scoops the buoyant eggs off the surface and transfers them to hatching troughs. It takes the eggs about three weeks to hatch at 42.8° F., and another six to seven weeks for the fry to metamorphose. Before metamorphosis the fry are fed on brine shrimp, otherwise known as *Artemia*; after metamorphosis, on chopped mussels and cheap fish. When a little over an inch long, the fry are shipped off to the growout facilities. One of these is a dammed-off sea loch at Ardtoe, Scotland, where work began in 1965.

At first the Ardtoe researchers were plagued by disaster. Heavy rainfall diluted the water in the loch, and many of the young plaice died from the lowered salinity. Plant life in the loch died, rotted, and used up the dissolved oxygen in the water. Predators and competitors for food flourished. However, by 1968 the culturists produced their first marketable crop. The original five-acre loch proved too large for management; so the fish are now reared in a variety of smaller enclosures in the loch and in the sea, including cages. The major obstacle is the low food conversion efficiency of plaice and sole—5 to 1.

More publicized is the project of Hunterston, Scotland, where hot effluent from a nuclear power plant is used to extend the flatfish growing season around the whole year (these fish, like most others, stop growing when the water falls below a certain critical temperature). By extending the growing season, the culturists at Hunterston produce plaice and sole of marketable size (at least 7.9 inches) in two years. In nature, it takes the fish three to four years to reach this size. However, it will probably be a long time until cultured flatfish make much of a splash on the maket, due to the high cost of feeding them. A cheap synthetic diet would make a great deal of difference.

CHAPTER **8**

The Peripatetic Eels and Other Minor Fish

Seldom eaten in the United States, eels are relished in many European and Asian countries. Broiled, fried, stewed, jellied, smoked, and pickled, the rich, firm meat of the eel is a delicacy. Eel is also easier to eat then most fish—eels lack the small, fine bones common in other fish and have only vertebrae. Germans and Middle Europeans dote on it. Tradition-minded Italian families feel that Christmas dinner is incomplete without an eel dish. Japan's gourmets have made such a dent in their regional eel fishery that Japanese fish dealers import eels by the planeload from the United States, where they are plentiful, and Japanese aquaculturists import millions of elvers (young eels) each year to fatten up to market size. Only one thing prevents eels from becoming a completely cultured crop from the egg on up, and that is the inability of aquaculturists to keep them alive through the very early larval stages.

The eel has one of the strangest life cycles of any fish. Unlike the anadromous fishes—salmon, shad, sturgeon, smelts, striped bass, and others—eels are born in the sea, spend most of their lives in fresh water, and return to the sea to spawn and die. This type of life cycle is termed *catadromous* (from the Greek *kata*, down, and *dromein*, to run).

99

Of the twenty-odd families of snake-shaped fishes that make up the order *Anguilliformes* (Latin for "eel-shaped ones"), only one, the *Anguillidae*, or freshwater eels, is catadromous. The rest are strictly marine fishes. But it is only the freshwater eel that is eaten to any extent by man.

Four species of eels are commercially important: the American eel, the European eel, the Japanese eel, and the Australian eel. American and European eels live in the Atlantic and connected seas. Japanese and Australian eels dwell in the Pacific. There is no discernible difference between American and European eels except that the European eels have seven to sixteen more vertebrae. Otherwise they are identical in size, behavior, and diet. Some scientists believe that they are actually one species.

The Japanese and Australian eels are quite similar to the American and European species. Japanese eels are believed to breed south of the island of Taiwan; Australian eels probably breed off Sumatra and Caledonia.

Both American and European eels spawn in the deep waters of the Sargasso Sea. It is believed that spawning takes place at a depth of about 1,500 feet, in absolute darkness and near-freezing cold. (Man has never observed it directly.) Each female eel releases from 10 to 20 million eggs, which float slowly to the warm surface layers of the water. There they quickly hatch into thin-bodied, transparent larvae called *leptocephali*. The leptocephalus (Greek for "puny head") resembles a long, thin willow leaf; it is mostly body, with a tiny, beaked head at one end. In the water, only two tiny black eye-spots ae visible.

The leptocephali swim feebly, probably feeding on plankton, and ride the currents of the Gulf stream, which will ferry them to their fresh-water destinations. The journey takes about a year for American eels and two to three years for European eels. However, the growth rates of the two species are so adjusted that they are at the same stage of development when they near land: the elver, two to three inches long and about as thick as a wooden matchstick.

The elvers undergo physiological changes that fit them for life in fresh water as they enter the mouths of rivers and streams. Still transparent at first, they soon turn brown, yellow, or greenish-black.

The males remain in the estuaries, where they grow to twelve to eighteen inches in length. The females travel far upstream to small tributaries, lakes, and ponds, often moving overland by night. Eels can survive out of water for hours as long as the sun does not dry them out, and their copious slime and tough skins protect them from injury as they slither across dew-wet fields. They can even climb a six-foot wall if it is uneven enough to give them a purchase. The females remain in fresh water anywhere from five to thirty five years, hiding by day and hunting by night. They reach a length of three to four or even five feet and may weigh up to fifteen pounds. Eels that take up residence in landlocked ponds may remain there all their lives, never leaving to spawn. Landlocked eels as old as fifty years have been taken. (An eel's age can be told by counting the rings in its otoliths, tiny, pebblelike concretions that form in the ear canals of fish and are believed to be part of their balancing systems.)

When eels approach spawning age, they turn silver in color and are known as silver eels. They make the best eating at this stage, as their flesh is saturated with fat to live on during the long journey of one thousand miles or more to the spawning grounds. It is believed that, like salmon, eels do not eat on the spawning run. In the fall of the year, spawning eels often form huge congregations in river mouths, where they are easily caught in nets.

The final migration to the spawning grounds probably takes place at a considerable depth, since no scientist or commercial fisherman has ever observed a mature eel swimming toward the Sargasso Sea nor taken one in their nets. Experiments indicate they may navigate by means of the earth's magnetic field,* although no magentic receptors have ever been pinpointed anywhere on an eel's body. After spawning, the adults simply vanish forever, probably sinking to the ocean floor and providing a meal to the creatures of the abyss.

In nature, eels are found even along the southern tip of Greenland, but in culture they are classed as warm-water fish, for they grow best between 77° and 82° F. Even so, it takes twelve months to get an eel from the elver stage to market size of at least five ounces, and this can be costly in terms of feed, because eels are not only

*It has also been suggested that eels may utilize faint electric fields created by ocean currents.

highly carnivorous but voracious, and such sloppy feeders that they waste about half of what is set up for them to eat.

In the wild, eels eat everything from carrion to smaller fish and even baby waterfowl. Years ago, before the Hudson River cleanup program began, there was a noisome pool below Schenectady into which several slaughterhouses poured their wastes. Almost nothing could survive in this polluted spot except eels, and they waxed huge and fat. In river mouths, eels make themselves very unpopular with fishermen by eating other species of fish caught in gill nets. In the Baltic region, eels have even been observed leaving the water to raid nearby bean patches.

However, in culture they are much more finicky, and won't even touch the food they spill. Eels are generally fed on chopped fish-market and cannery wastes, mixed with starch to make a doughy mass and fortified with vitamins. One investigator, Dr. John Poluhowich of the University of Bridgeport, Connecticut, has developed a low-cost eel diet that combines chopped earthworms, which he raises himself in a compost pile that provides heat for his eel tanks, and snap beans. He reports that eels also relish peas.

Scientists in Japan and Europe have been able to induce eels to spawn by means of hormone injections. However, the larvae always die in the first two weeks. Whether the scientists could not duplicate the precise water conditions they need or could not give them the right kind of food is not yet known. Once this barrier is broken, eel aquaculture will be truly established. Until then, culturists must depend on the yearly supply of wild elvers trapped as they ascend the rivers each spring. And this supply appears to be declining, due to overfishing, pollution, or both.

The leading eel-culturing nation is Japan, where eels have been raised for some 150 years. Most eels in Japan are raised in dirt ponds or enclosures with running water. The running-water enclosures yield up to an amazing 20,000 pounds per acre; the ponds yield about 1,100 pounds per acre, due to lower oxygen levels and high waste levels (the wastes are not flushed away). Total Japanese production is from 15,000 to 20,000 *tons* of eels a year. Some growers raise their eels in square concrete ponds, with rounded corners for easy cleaning, and some use raceways.

Cannibalism is always a problem with young eels, and they must be sorted out several times until they reach four to six inches, when they cease to be attractive to each other as food. The larger and smaller eels are placed in separate ponds each time. This practice also has the advantage of keeping the size of the eels in each batch more uniform.

Taiwan is second only to Japan in eel culture. Here the eels are raised in ponds whose sides are lined with masonry, concrete, or wood, to keep the eels from burrowing into them. The bottoms are left unlined so that the eels can bury themselves in the mud for their winter nap (they stop eating when the water temperature falls below 48° F.). The elvers were formerly fed on silkworm pupae, but the decline of the natural silk industry made it necessary to switch to ground-up trash fish and fish wastes.

The ground-up fish waste is formed into a thick, pasty loaf and placed in a wire basket open on one side. The eels literally burrow into the food as they greedily devour it. In Taiwan, eels are usually raised as part of a polyculture whose other members clean up after them: various carp to dispose of the spilled eel food and other organisms that it nourishes, and mullet to eat up the blue-green algae that thrive on the high concentration of nitrogenous wastes in the water. A synthetic, dry ration that contains over 50 percent of fish meal and smaller amounts of starch and soybean powder is also used.

West Germany and Russia have commercial-scale eel farms. In the United States, eel farming has generally been unprofitable due to the difficulty of marketing the product. But the growth of foreign markets may change the situation. The New Jersey utility PSE & G grows eels in a heated tank inside a plastic greenhouse. In North Carolina, a research team from the Univesity of North Carolina has been raising eels intensively using Japanese techniques. In Virginia, on Chesapeake Bay, a private culturist raises eels for export to Europe, and across the bay in Maryland an American subsidiary of a Japanese corporation is raising them for the Japanese market. Normally, the Japanese consider American eels second-rate and will not pay top prices for them, but apparently the cultured eels are equal in quality to Japanese eels. Japanese consumers prefer eels from ⅓ to ⅔ pound in weight; Europeans like them to weigh at least ⅔ pound. However, these are minor matters.

Several species of fish that are best known as small aquarium specimens in the Western world are raised as food fish in Southeast Asia. One is the giant gourami, which grows to two feet in length and lives on a vegetarian diet of plankton and leaves. The smaller kissing gourami, an aquarists' favorite, grows only to ten inches, but is also esteemed for its flesh. Several other species of gouramis are raised for food in Southeast Asia, often in polycultures with carp.

The snakehead, another tropical fish, is raised in Taiwan and Southeast Asia. Snakeheads grow up to three feet long and, like the gouramis, can breathe air for long periods. They are carnivorous and used to be considered pests in fish ponds. However, their flesh is excellent and it was discovered that they could play a useful role as predators in carp and tilapia ponds. The snakeheads prey on un-wanted wild fish and also thin out the excess population of carp or tilapia fry. Harvested along with the main crop, they sell at a very profitable price.

One of the oddest fishes raised commercially is the mudskipper, a tropical saltwater fish that is raised in Taiwan. Mudskippers are small fish, reaching only about five to six inches maximum length, but they are esteemed by Chinese gourmets. They are named for their habit of leaving the water at low tide and hopping around the mud flats on their pectoral fins. They live mainly on algae, which they graze off the bottom.

Once abundant but now becoming scarce are the sturgeons, a family of primitive fishes with heavily armored bodies and long, shovel-shaped snouts. There are about two dozen species distrib-uted throughout the cool waters of the Northern Hemisphere. Most are anadromous; a few are strictly freshwater. The giant of the sturgeon clan is the beluga, native to the Black and Caspian Seas and the Volga River. This ponderous fish may reach a length of more than twenty feet. The largest American sturgeons are the white sturgeon, of the Pacific Northwest, which grows to fifteen feet and may weigh over one thousand pounds, and the Atlantic sturgeon, whose record size is fourteen feet and eight hundred-odd pounds. The pigmy of the group is the sterlet, a European species that grows only two feet long. The big sturgeons are extremely long-lived. Some specimens have been estimated at 150 years of age.

Sturgeons swim slowly along the bottom, vacuuming up small animals and plant matter with their toothless mouths. Most species have gizzardlike stomachs with which they crunch up snails and clams. Sturgeons spawn in shallow, clear streams with gravel beds, where they are very vulnerable to fishermen.

Sturgeons are famed for their roe, the high-priced delicacy we know as caviar. Their flesh is also delicious, and their swim bladders yield a high-quality gelatin called isinglass.

Overfishing, pollution, and blocking of spawning streams by dams have lowered the world sturgeon population. Unfortunately, the only way to extract caviar from a pregnant sturgeon is to kill her and slit her belly open.

Hatchery programs for restocking the sturgeon fisheries were begun in the late nineteenth century in both Russia and the United States. The United States programs were eventually dropped due to the difficulty of keeping the newly hatched fry alive and of capturing ripe males and females at the same time. Due to their size and stiff, armored bodies, it is not possible to strip sturgeon by hand, as with trout.

Russia, which produces over 90 percent of all the world's sturgeon and caviar, has the only major hatchery program in operation today. Mature fish (at least fifteen years old) are netted and given hormone injections, then held in tanks until they are judged to be ripe. The fish are then killed and their reproductive organs removed for fertilizing the eggs. Although sturgeons are large, their eggs are small, and they hatch in four to ten days, depending on the species and the temperature of the water.

The sac fry are put in shallow troughs or basins with running water and fed a special mixture of *Daphnia* and another tiny freshwater crustaceans. The fry grow rapidly and after ten days are big enough to put into ponds, where they are fed on *Daphnia* and bristleworms for one to three months. When they are three to four inches long the young sturgeons are transported to river estuaries, where they are released. Formerly they were released upstream at the hatcheries, but it was learned that most of the sturgeon fingerlings were gobbled up by predators on their way to salt water.

Since sturgeons grow extremely slowly, they are not suitable for true aquaculture. However, the sterlet reaches market size (four to five pounds) in five or six years, and the high price it commands makes up for the lengthy grow-out period. It also eats low on the food chain, living mainly on insects and their larvae. Sterlets are raised along with carp in Hungary.

A note of reassurance to caviar lovers: An expatriate Russian fisheries biologist is working in California on a method to remove caviar surgically without harming the female sturgeon. If perfected, this method could lead to culture of sturgeons for caviar.

CHAPTER **9**

The Cultured Crustaceans

"Shrimp boats are coming!" began a hit song of the 1950's. The shrimp fleets are still making profitable hauls, but the chances are strong that within the next few decades a large share of the world's shrimp will be cultured. The same holds true for many of the other popular shellfish.

"Shellfish" is a misnomer, a hangover from the days when uncritical Anglo-Saxons called anything that lived in the water a "fish." Shellfish are not fish at all. The term covers two huge and unrelated groups of animals: the crustaceans, such as shrimp, crabs, and lobsters, and the mollusks, such as oysters, clams, and scallops. The crustaceans belong to the phylum *Arthropoda* and are thus related to insects, spiders, and scorpions. In some cases these relationships are easy to see. Some crabs look unappetizingly like spiders, and scorpions look very much like lobsters. However, insects have six legs, spiders and scorpions have eight, and the crustaceans eaten by man have ten legs.

Like their arthropod relatives, the crustaceans have segmented bodies encased in a shell that serves as a skeleton and sometimes as armor. Such an external skeleton is called an exoskeleton, and the creature that wears one can grow only by molting its exoskeleton

107

periodically. This has important consequences for the animal's life cycle.

The first crustaceans appeared on earth perhaps as long as 600 million years ago, as scientists interpret the fossil record. This makes them considerably senior to the fishes, which made their first appearance nearly 200 million years later. Unlike fishes, whose newly hatched young bear at least a faint resemblance to the adult form, the crustaceans go through a lengthy series of metamorphoses before they assume their adult shapes. Some of the early stages are absolutely grotesque looking, and they are tagged with a bewildering collection of names: nauplius, zoea, megalops, mysis, phyllosoma, puerulus, and so on ad infinitum, or so it seems to the layman. Adding to the confusion, biologists do not apply the names uniformly.

Some of these stages are peculiar to certain species, and some species skip an entire set of stages, which makes the whole business even harder to understand, even for biologists. From the viewpoint of the culturist, this multiplicity of metamorphic stages becomes a nightmare, because each stage requires a different diet, and in many cases the requirements still elude identification.

Crustaceans are much more difficult to raise than fish and mollusks. The metamorphic cycle is only one of several problems. Another is that most of the crustaceans are voracious cannibals at one stage or another of their lives, and some are cannibalistic all the time. They are, of course, particularly vulnerable when they are molting and their new shells have not yet hardened. Bottom-dwelling crustaceans, such as lobsters, tend to be highly territorial, which does no harm in the ocean but can cause fatal conflicts in the cramped territory of a tank or pond.

Economic as well as biological factors enter in. Most of the crustaceans are voracious carnivores, which means that it is very expensive to feed them. In addition, the food conversion efficienty of some species is rather poor. Lobsters, for instance, need fifteen pounds of food to make one pound of lobster. (Scientists say that one reason for this inefficiency is the tremendous amount of energy used up in molting and secreting a massive new shell.) Shrimp, however, with their thinner shells, have conversion ratios as high as 2.2 to 1. Then, too, much of the biomass of a crustacean is waste: shells, heads, gills, etc. And in many species we consume only a part of the animal. With shrimp, for example, we eat only the tail.*

Since we have begun this chapter with shrimp, let us take the brown shrimp as an illustration. This is a commercially important species found in the Gulf of Mexico and off the east coast of the United States, where it ranges as far north as Cape Cod in the warmer months. Brown shrimp grow to seven or eight inches if they live long enough, and about half that length is edible.

Brown shrimp mate in a rather fastidious way after the female has molted. Shrimps are gregarious creatures, and apparently the newly molted female releases a pheromone that attracts nearby males. The same phenomenon seems to hold true for other crustaceans, even such solitary ones as lobsters.

With his claws, the impassioned male shrimp places packets of sperm in a special receptacle on the female's belly. The female carries the spermatophores with her until her eggs are mature enough for fertilizing. She then spawns in a solitary ritual, always at night and usually around midnight. Burrowing into the sand or mud of the bottom, she releases a cloud of sperm and eggs, tens or hundreds of thousands of them. The fertilized eggs drift off to continue their predestined development into adult shrimp, although only one in ten thousand even survives the larval stages.

The eggs hatch in about twelve hours into bizarre little animalcules called *nauplii*, the first stage of larval development for most crustaceans. The nauplius has a pear-shaped body about 3/10 millimeter long, a mere speck to the unaided human eye. Each nauplius possesses a single eye and three pairs of swimming legs, armed with bristles that give them a feeble purchase on the water. Living on stored yolk, the nauplii molt four times within the next forty-eight hours, roughly doubling their size. At the fourth molt they metamorphose into a new shape called a *protozoea*, which has a long tail and an enlarged head sporting a pair of protuberant eyes and a defensive growth of spines and bristles.

Aggressive and active, the protozoea begin feeding on single-celled algae and perhaps also on organic detritus. Passing through

*However, a shrimp expert points out that the tail meat is about 60 percent of the animal's weight.

three stages, they reach a length of three millimeters and molt again at the age of six days to pass to the next stage, the *mysis*. The mysis sports a fearsome, beaklike spine jutting forward between its stalked eyes. In some species of shrimp it is a simple lance; in others, it zigzags like a conventionalized lightning bolt. The mysis begins to look like an adult shrimp as it acquires all ten legs, with tiny, functional claws at their ends, the characteristic long abdomen of shrimp, and a broad, fanlike tail that whisks it away from danger with a single powerful snap. The mysis itself switches from plant to animal food. About two weeks after spawning the mysis makes its final metamorphosis, into the post-larval stage. After three more molts, the shrimp assumes the fully adult form and grows rapidly, from 25 to 45 millimeters (1 to 1¾ inches) a month.

Although spawning takes place in deep water, by the end of the mysis period the young shrimp have somehow made their way back to the coastal shallows and the brackish estuaries of the rivers and creeks. It is not certain how they manage this trip. One theory is that they swim along with the incoming tidal surges and cling to seaweed and other objects on the bottom while the tide flows out. In this way they could move slowly but surely a few miles each day. The young shrimp spend five to seven months in the estuaries and coastal waters until they have reached half their adult size, then depart once more for deep water. There they spend the rest of their lives, hunting close to the bottom by night and hiding in the mud or sand by day.

The details of crustacean development and life cycles differ even within the same genus; so we shall draw a veil over the confusing scene and proceed to other matters.

Brown shrimp belong to the genus *Penaeus*, to which most of the commercially important shrimp belong. The penaeids range in length from the thirteen-inch *sugpo* of the Philippines to a member of one- and two-inch species. The most important commercial shrimp species in the United Staes are the brown, white, and pink shrimps, all penaeids. Most of the research on shrimp culture involves penaeids.

Shrimp have been cultured in Southeast Asia for at least five hundred years, using wild larvae. The first cultured shrimp were undoubtedly strays that entered fish-culture ponds, grew to market size there, and were harvested along with the fish. Today shrimp ponds are built so that the incoming tide sweeps in new batches of shrimplings. At the peak of high tide the culturist puts a screen over the sluice gate to keep any shrimps from swimming out. Since the pond contains shrimps in all stages of growth, harvesting goes on year round. The shrimp are netted at night as they head for the sea on an ebb tide.

In traditional southeast Asian shrimp culture, little is done by man except perhaps to get rid of some of the numerous predators. The shrimp forage for themselves on the natural food chain. Even so, a well-tended pond can yield as high as half a ton of shrimp per acre a year.

In the Philippines the culture of a big penaeid shrimp called *sugpo* is a thriving business. Sugpo can reach a weight of a quarter pound, although they are usually harvested and sold at half that weight or less. It takes sugpo six to eight months to reach market size.

Philippine shrimp culture is more methodical than that of the rest of Southeast Asia. Many farms have nursery ponds in which the young shrimp, captured on bundles of grass or twigs set out in the estuaries, are fattened up on lab-lab before being placed in the grow-out pond. The more progressive sugpo growers fertilize their ponds to boost the natural food productivity as well as feeding the shrimp on trash fish or meat scraps. Those growers who can afford an aerator find that it increases their yields as much as threefold. The shrimp are harvested when they are 4¾ to 6 inches long, after six to eight months. They are often raised together with milkfish.

Sugpo are also raised in Taiwan,* where spawning in hatcheries supplements the supply of wild seed shrimp. Since it is difficult to raise sugpo to maturity in captivity, many spawners are still obtained from fishermen. They can be recognized by their swollen ovaries, which show up clearly as dark spots under the shell of the shrimp's back.

The spawners are placed in plastic or concrete tanks. To prevent

*In Taiwan they are called grass shrimp or tiger prawns.

overcrowding of eggs and larvae, the spawners are moved from tank to tank, discharging only a part of their cargo of eggs in each. (A sugpo spawner releases more than 300,000 eggs.) The eggs hatch in about thirteen hours, and the nauplii transform themselves into zoea at fifty hours. For three to four days they live on cultured one-celled algae, then enter the mysis phase and feed on zooplankton, which is gathered from milkfish ponds with fine-meshed plankton nets. After a few more days they are switched to a diet of ground shellfish meat, fish, or egg custard. At a length of about 4/5 inch they are sold to the growers for planting out. In Taiwan, sugpo are almost entirely raised in milkfish ponds, where they thrive on the natural food chain.

Taiwanese shrimp farmers have a problem of climate. Whether they get their fry from hatcheries or fishermen, they get them too late in the year to raise them to harvest size in one season. Since sugpo are tropical animals and quite sensitive to cold, they must be carried through the winter in special sheltered ponds. They are kept semi-starved during the cold season to keep them small, which avoids the problems of too much biomass for the volume of the pond.

The most advanced shrimp farming is done in Japan, where the giant ten-inch penaeid shrimps called *kuruma-ebi* command a fantastic price in restaurants. Kuruma-ebi means "wheel shrimp"; when curled up, the kuruma resembles a spoked wheel. It is the prime ingredient of the best *tempura*, a mixture of seafood and vegetables dipped in batter and deep-fried in oil.

Kuruma culture in Japan is the brainchild of a dogged scientist named Motosaku Fujinaga. Dr. Fujinaga was first able to breed kuruma in captivity as far back as 1934, but he was not able to bring them past the larval stages until 1942. Success came when he discovered that the larval shrimp had to have food right in front of them. Food lying on the bottom of the tank or floating through the water even a few inches away was ignored by the larvae. By developing a process that kept the algal culture stirred up and evenly distributed throughout the tank, he enabled the kuruma larvae to feed and survive.

As a labor-saving device, he cultured algae directly in the tanks where his larval shrimp darted about, rather than following the standard practice of growing the algae separately.

112

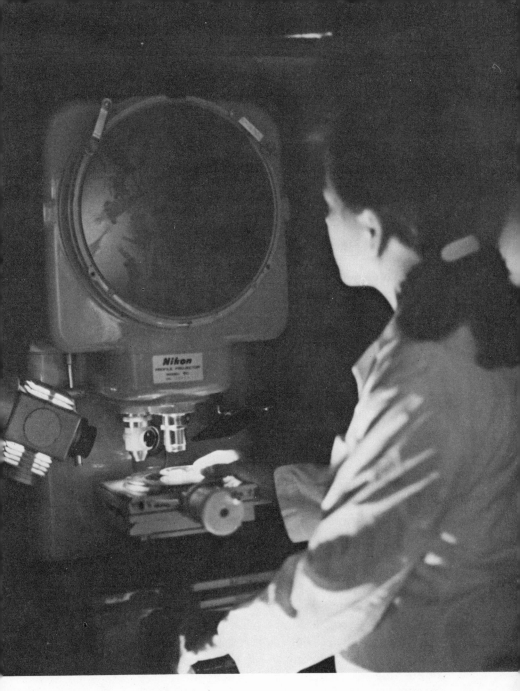

*In a Taiwanese laboratory, a scientist studies larval shrimp under very
high magnification.*

At the mysis stage, Fujinaga switched his shrimp to a diet consisting mainly of brine shrimp, which at one point he had to import from California. As post-larvae, the shrimp are given a ration of chopped worms, bivalves, and fish. At this stage they also become highly cannibalistic unless fed to repletion. To prevent fouling of the water, Fujinaga invented a double-bottomed tank whose inner bottom is a fine-meshed screen. On top of the screen goes two inches of sand. Air is pumped up through the screen and the sand, providing ample oxygen. It also sets up a circulation pattern that brings the water down through the sand, which acts as a filter. The sand fills a double function by giving the shrimp a place to hide during the day, as they would in the wild. In fact, all kuruma grow-out ponds, no matter how simple the design, have a sand layer on the bottom.

This habit of burying themselves in the bottom makes kuruma harvesting difficult. Good results have been attained by using a high-pressure hose mounted on the trawl net to blast the shrimps out of the sand. Even better results have been obtained by using a mild electric shock to flush the shrimps out of hiding.

Once harvested, the kuruma are chilled to 55° F. and packed carefully in dry sawdust, which insulates them. With their metabolism barely functioning at this temperature, the shrimp can survive up to four days in winter or two days in summer, living on the oxygen contained in the water trapped in their gill cavities. Not just any sawdust will do: it must come from the cryptomeria, or Japanese cedar, whose wood has natural insect-repelling qualities but does not harm the shrimp nor give them an odor. At the restaurant, the kuruma are kept in tanks of water until required for a meal. They are dipped out, beheaded, peeled, and cooked on the spot, ending their carnivorous lives amid gastronomic pomp and ceremony.

Japanese kuruma-growing methods are now being applied in Hawaii, where aquaculture is a fast-growing industry. Elsewhere in the United States, shrimp culture is still in the experimental stages. The Coca-Cola Company and F.H. Prince & Company sponsor an interesting self-contained project in Mexico, where a team headed by a University of Arizona scientist, Dr. Carl Hodges, raises shrimp in a giant plastic greenhouse. The shrimp are fed a synthetic ration that is based on wheat. Algae in the raceways provide oxygen as well

Scientists at the University of Arizona's Environmental Research Laboratory raise blue shrimp (Peneaus stylirostris), a Pacific species, in an environmentally controlled greenhouse at Puerto Peñasco, Mexico. The entire reproductive cycle takes place in the culture facility.

UNIVERSITY OF ARIZONA ENVIRONMENTAL RESEARCH LABORATORY

as a dietary supplement. Waste water from the shrimp raceways, full of algae, goes to other ponds to feed mullet, a salable byproduct.

Dr. Hodges has developed techniques for inducing his captive shrimp to breed at any time of the year. This not only assures a year-round supply of shrimp but also makes it possible to start a selective breeding program. The Arizona scientist sees his success with his pilot plant on the Gulf of California as a model for the productive use of desert coasts, of which the world has about twenty thousand miles.

115

The ability to manipulate the spawning of shrimp is spreading. Recently researchers at the National Marine Fisheries Service lab in Galveston, Texas, succeeded in getting the shrimp to mate and spawn by controlling their diet, day length, and water temperature and salinity. British scientists have taken sugpo through several generations in a closed-cycle experimental hatchery. In France, researchers are breeding kuruma experimentally. It seems only a matter of time before shrimp hatcheries are supplying stock to growers throughout the warmer regions of the world and to thermal aquaculturists in colder areas.

One of the glamour creatures of curstacean culture is the giant freshwater prawn *Macrobrachium rosenbergii.** Native to Southeast Asia, *Macrobrachium* males can reach 10 inches in length (females seldom exceed six inches, and there is a high proportion of runts among males). *Macrobrachium* are handsome animals, resembling a light buff lobster with deep blue claws. There, however, the resemblance ends. Instead of the heavy claws of the lobster, the *Macrobrachium* has exceedingly long slender claws nearly as long as its body.

M. rosenbergii have many advantages for aquaculture. They adapt well to captivity. They can stand a wide range of temperatures (from 59° to 95° F.) and salinities. The females retain their fertilized eggs until they hatch, which considerably reduces mortality. The vulnerable larval stages are relatively short. And the young prawns grow fast, reaching market size in seven to eight months. Their food conversion ratio is relatively good: 3.3 to 1. Last but not least, the meat is excellent.

In the wild, *Macrobrachium rosenbergii* spends most of its life in fresh or brackish water. It spawns in the brackish water of river estuaries. The male deposits his sperm on the female's legs, and she spreads it over the eggs, which she keeps in special brood chambers under her abdomen. In about nineteen days the eggs hatch, and the tiny larvae are carried out to sea by the current. Their life cycle requires that they spend their larval stages in salt water, where they

*Neither scientists nor lexicographers can agree on what distinguishes a "shrimp" from a "prawn." Some use "shrimp" for the smaller species and "prawn" for the larger ones; others call the saltwater species shrimp and freshwater species prawns.

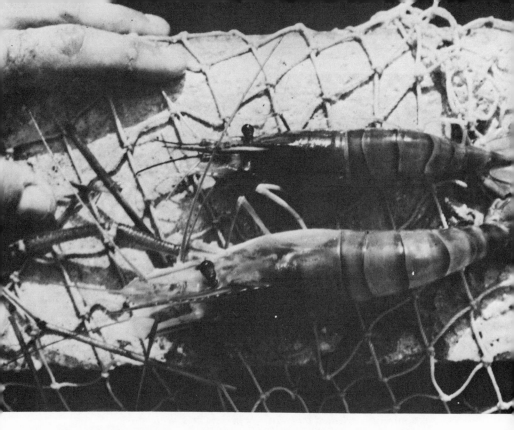

A male and a female Macrobrachium rosenbergii *side by side. The male (lower) is much larger and has strikingly larger claws than the female.*

feed mainly on zooplankton. After eleven molts they metamorphose into juveniles and start crawling and swimming slowly back into the river mouths. Some remain in brackish water; others travel as far as forty miles or more upstream, molting frequently as they grow. From the time they leave the sea, they will eat almost anything, living or dead, including plant matter and their small brethren.

Long considered a delicacy in southern Asia, *Macrobrachium* were traditionally caught in traps and on hook and line. Another technique used the poacher's trick of jacklighting to temporarily immobilize the prawns while they were scooped up with a hand net. A small-scale cultural program was begun in Malaysia, but the *Macrobrachium* industry really owes its origin to a Hawaiian en-

117

thusiast, Dr. Takaji Fujimura of the Anuenue Fisheries Research Center on the island of Oahu.

Dr. Fujimura learned that an FAO scientist named Shao-Wen Ling had succeeded in breeding *M. rosenbergii* in captivity at the Malaysian government fisheries lab at Penang. In 1965 he imported thirty-six of the big freshwater prawns from Malaysia. They became the parent stock of millions of baby prawns that the state-owned hatchery at Anuenue distributed to culturists in the next twelve years.

In addition to developing a breeding program, the Anuenue scientists worked out improved pond designs and grow-out techniques, learning as they went. The first demonstration ponds were stocked with hatchery-bred juvenile *Macrobrachium* in 1967. The first commercial operation, with sixteen acres of ponds, began in 1972. By 1977 Hawaii had seventeen *Macrobrachium* farms, with an average production of three thousand pounds per acre per year.

Taiwanese scientists examine a good-sized specimen of Macrobrachium rosenbergii, *relished by gourmets in Asia and elsewhere.*

CHINESE CULTURAL CENTER

The prawns are stocked in the grow-out ponds at about ⅓ inch in length. The farmers try to keep a rich, green bloom of algae in the ponds to use up the wastes given off by the prawns and also contribute to the food chain of the ponds. Harvesting begins about seven months after stocking. It is done with a specially designed net whose two-inch meshes let undersized prawns slip through without their appendages becoming entangled. Five inches is the minimum size that is caught. Due to the great difference in growth rates between "bulls" (big males) and runts, the harvest season can be extended for many months.

A good deal of experimental work is being done with *Macrobrachium* in the Gulf state, where it is being seriously considered as a warm-season crop in conjunction with trout as a winter crop in the same ponds. So far these endeavors, while biologically satisfactory, have foundered on the problem of labor costs.

Macrobrachium rosenbergii have also been considered for culture in raceways heated by power-plant effluent. To prevent cannibalism, the prawns are kept in "condominiums"—stacks of plastic cages each with its own little feeding rack outside. PSE & G, the New Jersey utility, carried on such a program for several years with good growth rates. However, with the shrimp condominiums they could not get high enough densities for a profitable harvest, and without them cannibalism set in, destroying the harvest anyway. These problems will probably be solved in the future, making high-density culture possible.

There are numerous species of *Macrobrachium* besides *rosenbergii*. One, *M. ohione*, is native to the United States, where it is known as river shrimp. These species may one day also be used in culture.

The most successfully cultured crustacean in the United States at present is the crayfish, a freshwater denizen that looks like a midget lobster and is found from Mexico up into Canada. Indeed, these hardy and prolific little animals are native to every continent save Africa and Antarctic.

People quarrel violently over whether the proper name is crayfish or crawfish. Either form is correct, but the specialists seem to prefer

A promotional photograph features Macrobrachium *in a decorative culinary context. Marketing is essential to the success of any aquaculture business.*

COURTSEY OF THE WEYERHAEUSER COMPANY

crayfish. The word, by the way, comes from the Old French name, *crevice,* which comes in turn from an even older German word for crab.

In North America there are some 250 species and subspecies of crayfish (classification is not yet completed), with twenty-nine species in Louisiana alone. Thanks to Creole and Cajun cuisine, Louisiana is the center of world crayfish culture. About 90 percent of our commercial crayfish production, including the very large harvest of wild crayfish, comes from that state, from twenty to twenty-five million pounds a year. About 85 percent is consumed right there. Oregon ranks second in crayfish production, selling much of its harvest to California under the name of "short lobster."

Crayfish range in size from the inch-long dwarf crayfish of Louisiana to a giant Australian species that reaches sixteen inches in length and weighs up to eight pounds. Most species are burrowers; this habit causes them to regarded as pests in many parts of the world, for their burrows cause leaks in dikes and levees, undermine man-made ponds, and cause general havoc. In addition, crayfish eat the succulent young shoots of rice plants in the paddies. Although crayfish prefer meat, they are true omnivores.

Crayfish of one species or another can be found in almost every kind of freshwater habitat, from fast-flowing streams to sluggish creeks and swamps. Crayfish have been trapped in deep lakes at a depth of 104 feet; others thrive in roadside ditches and puddles. Rice fields provide almost ideal growing conditions for crayfish.

Crayfish farming is thought to have begun through an accidental discovery by a Louisiana rice farmer. According to tradition, one fall this man flooded his rice fields to attract ducks for hunting. The following spring he found a bumper crop of crayfish. The technique has changed little since then, for crayfish thrive on a minimum of care. Although crayfish culture got off to a slow start because of the abundant supply of easily trapped wild crayfish, by the late 1970's there were over 50,000 acres in crayfish production in Louisiana alone, with additional acreage in other states.

Two species of crayfish are cultured in Louisiana: the red swamp crayfish (*Procambarus clarki*) and the white river crayfish (*Procambarus acutus acutus*). The two species are similar in size and probably in life history. In early summer (May or early June in Louisiana) the female crayfish comes out on shore and digs a burrow down to the water table. Burrows can go as deep as forty inches, although most are shallower. She retreats to the bottom of the burrow, followed by a male. Mating takes place in the burrow or in an open pond: The male deposits sperm in a receptacle on the female's body, where it is stored until the eggs are laid in late summer or early fall. The eggs are fertilized as they are laid, and are attached to the underside of the female's tail with a sticky secretion. Hatching takes place two to three weeks later. The young remain on their mother for another week while they undergo two molts. About this time, rising water floods the burrow and softens the clay plug with which the female has barricaded the entrance. The female crawls to the surface with her young still attached to her tail, but they soon break free and begin life on their own. Red and white crayfish live one to two years. Many females die shortly after hatching; others survive, molt again, and are plump and meaty when caught in November.

Louisiana crayfish are raised in rice fields in rotation with the rice, in open ponds constructed solely for crayfish, and in ponds built in wooded swampland. Ideally, the pond should be twelve to thirty inches deep and in the open. Trees and bushes interfere with the harvest. They also shade the water, inhibiting the growth of water plants on which crayfish feed, and keeping the water cool (crayfish like it warm). The leaves they shed into the water use up oxygen as they decay. Crayfish can survive in quite foul water by climbing out to get air, but their growth is slowed significantly.

Crayfish ponds range from a few acres to over three hundred acres. The largest at this time covers five thousand acres, but experts recommend twenty to forty acres. The water should be hard and slightly alkaline to assure sturdy shells, an important survival factor. Under favorable conditions—a warm fall and plenty of food—young crayfish hatched in mid-September can reach market size (at least three inches) by mid-December. Harvesting continues through the winters. The main harvest occurs in spring, from March to May. Yields in a well-managed open pond may run as high as eighteen hundred pounds per acre a year.

Because of their bottom-dwelling and burrowing habits, cultured crayfish are harvested in traps and baited lift nets, like their wild relatives. Contrary to popular belief, the bait works better the fresher it is. The favorite bait is fresh chunks of gizzard shad or fish heads. Chicken necks are also effective and cheap. In an emergency, a perforated can of dog food can be used. If traps are used in low-oxygen water, the crayfish may be asphyxiated if left too long. This renders them unfit for food, but it can be prevented by leaving a corner of the trap sticking up out of the water so that the crayfish can climb up to breathe.

The ponds are drained in May and June, when the crayfish become tough and the harvest comes to an end. The water is let out slowly to give the crayfish ample time to dig burrows. Without this protection the loss to raccoons, birds, and other predators would be very high. The ponds remain dry for four to ten weeks, which eliminates most water-dwelling predators and gives the farmers a chance to eradicate troublesome water plants. The ponds are flooded again in September.

Unlike the trout farmer, the crayfish grower has no feed bills. About 80 percent of the crayfishes' diet is naturally growing vegetation. The remainder is mostly small, slow-moving animals such as worms, snails, and larvae, which are also part of the natural food chain. Stock crayfish are needed only for the first year of a new pond, since enough crayfish remain untrapped in established ponds to renew the population. Stocking is done inMay and early June, when the animals are tough and prices are at rock bottom.

About 85 percent of a crayfish is waste. Scientists at Louisiana State University have found that this waste (heads, legs, and exoskeleton) can be ground up to produce a highly nutritious catfish chow. Due to the seasonal nature of the crayfish harvest, however, this process is not yet commercial. Other, still experimental, uses of crayfish are for control of water weeds and in polyculture with fish. One complex scheme involves growing crayfish in watercress ponds to provide a salable by-product.

In many states crayfish are raised as bait for sport fishing. Nothing, it is said, attracts a smallmouth bass like a freshly molted young crayfish.

In much of Europe the crayfish is highly esteemed. In the Scandinavian countries the brief crayfish season at the end of summer is a time of parties and festivities. In former days each guest was supposed to eat at least twenty crayfish and drink a small glass of aquavit with each claw. Today, however, overfishing and a disastrous fungus plague have created a severe scarcity.

One solution has been to import crayfish from Russia and Turkey. Restocking crayfish waters with disease-resistant American species is also being done. A hardy species from the Mississippi Valley, *Orconectes limosus*, stocked in France and Poland, has become well-established. European gourmets complain, however, that this crayfish is smaller than the European type and not as tasty. But a West Coast species, *Pacifastacus leniusculus*, has turned out to grow faster than the European crayfish and to be every bit as good. A hatchery in the south of Sweden now produces young *Pacifastacus* for stocking in twelve countries.

The most prized of all crustaceans is the lobster, which until recently has been also the most difficult to culture. True lobsters belong to the genus *Homarus*, and two species are commercially important. These are the European lobster, *H. gammarus*, and the American lobster, *H. americanus*. The two species are very similar and will hybridize freely in captivity. The main difference is that the European lobster is slightly longer than the American at each stage of growth.

Denizens of the Atlantic Ocean, these lobsters live in water that ranges from just above freezing to the seventies. In the lab, they can survive temperatures into the nineties with the aid of massive oxygenation, but they are under such physiological stress that they will not grow.

Lobsters inhabit—or used to—every depth from the low-water mark to 3,500 feet or deeper. However, the rampant overfishing that has made them scarce has almost eliminated them from the shallow waters offshore. Solitary, aggressive, and cannibalistic, lobsters need shelter such as rocks or shipwrecks. They are highly territorial and will sometimes fight to the death in defense of their territories, but a repertory of ritual threat gestures usually averts actual combat.

Some lobsters spend the winter in deep water, then move inshore to warmer water in summer to mate. Mating occurs when the female molts. Like other crustaceans, she gives off a pheromone at this time that attracts male lobsters. The male, drawn by the scent, approaches its source cautiously. Tiptoeing over the bottom, he moves his antennae rapidly from side to side to zero in on the female and continually opens and closes his jaws. Why he does this is not known, but researchers are much impressed by the display. On reaching the female, who is rather sluggish after the effort of escaping from her old shell, the male caresses her all over with his antennae. She responds with apparent moderate enthusiasm. After about half an hour of this antenna play, the male gently rolls her over onto her back, mounts her, and inseminates her. As with the other crustaceans, the sperm remains in a special receptacle until needed. Mating is most successful when the male and female lobsters are about the same size. A small male can inseminate a big female, but a large male and a small female are anatomically mismatched.

Mating usually takes place within a few hours after the female has cast off her old shell, and if it doesn't happen within a day or two, it is no go. Once the female's shell has hardened, the loses her sexual receptivity and will attack the male as an intruder on her territory. For that matter, it is only the influence of the reproductive pheromone that keeps the male from gobbling up the newly molted female as she lies defenseless in her den.

From nine to thirteen months after mating, the female extrudes her eggs, fertilizing them at that time with the stored sperm. Prior to this process, the female lobster spends several days in grooming her abdomen and swimmerets. Retiring to a secluded place, she assumes a tripod stance on her two big claws and the tip of her tail. Then she painstakingly goes over her abdomen and swimmerets with her bristly rearmost pair of walking legs. Only when every scrap of dirt is brushed or picked off does she begin to lay her eggs.

The female turns over on her back and curls her tail around toward her head, forming a sort of cup. She rakes the fertilized eggs together as they emerge in thin strings from her twin ducts and guides them toward their resting place. From 5,000 to 125,000 eggs are laid, depending on the age and size of the lobster, and probably also the food supply during the intervening period. The eggs are cemented to the hairs of the swimmerets on the underside of the female's tail, where they incubate for ten to twelve months. Thus as much as two years may elapse between mating and hatching. This slow reproductive rate is another factor in the decline of the lobster.

A female carrying eggs is said to be "berried." The eggs, about the size of a tomato seed, are greenish-black when first laid, then turn green and eventually brown as cell division continues. The eggs usually hatch when the water temperature reaches 50° to 68° F. in spring or early summer. Although the long gestation period is a disadvantage in one way, in another it is advantageous because the lobster larvae pass through the extremely vulnerable nauplius and protozoea stages in the egg, emerging as full-fledged mysis, with a much better chance of survival. Even so, it is estimated that only one larva in ten thousand reaches the post-larval stage.

A berried female lobster, carrying a large crop of eggs beneath her abdomen. Wooden pegs in each claw keep the claws shut, a necessary precaution in handling these aggressive animals.

PHOTO BY GARETH W. COFFIN, MAINE DEPARTMENT OF MARINE RESOURSES

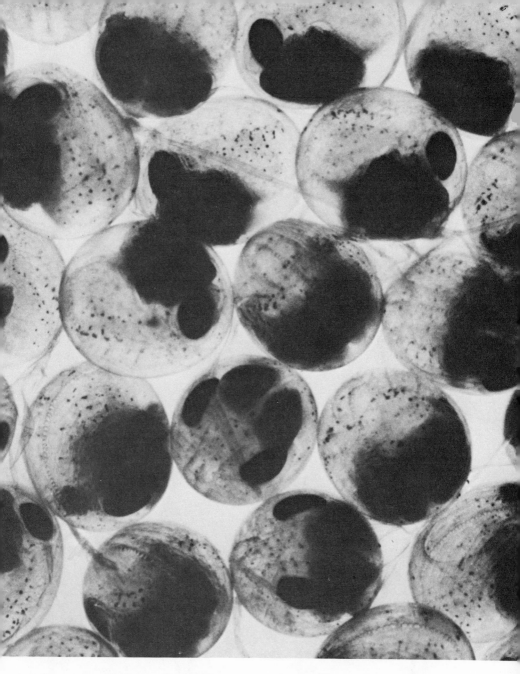

Lobster eggs almost ready to hatch. Dark eyespots and yolk sacs show up prominently. Actual size of eggs is about 1/16 inch.

PHOTO BY GARETH W. COFFIN, MAINE DEPARTMENT OF NATURAL RESOURSES

LATERAL VIEW

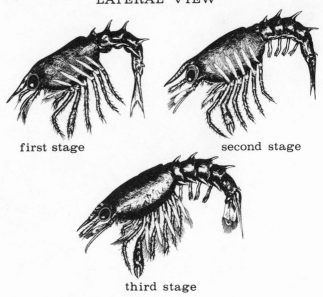

first stage   second stage

third stage

Stages of development of the lobster larva.

FROM FRANCIS H. HERRICK, *The American Lobster.*
COURTESY OF MAINE DEPARTMENT OF MARINE RESOURSES

DORSAL VIEW

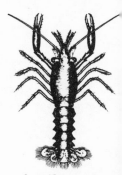

first stage   second stage   third stage   fourth stage

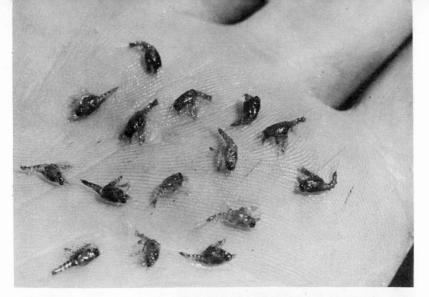

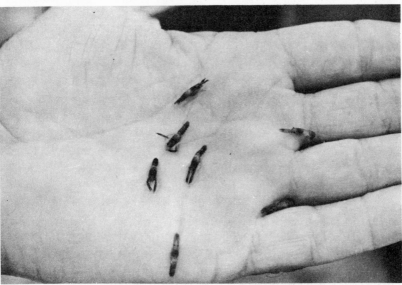

Top: *First-stage lobster larvae, about eight days old, bear little resembl-ence to the adult form. Man's hand gives an idea of their size.*
PHOTO BY GARETH W. COFFIN, MAINE DEPARTMENT OF MARINE RESOURSES

Bottom: *Postlarval lobsters, probably fifth-stage (after the fourth molt). Two seem to be backing into "dens" between the biologist's fingers. Seeking concealment is instinctive for postlarval and adult lobsters and is an important survival factor.*
PHOTO BY GARETH W. COFFIN, MAINE DEPARTMENT OF MARINE RESOURSES

The newly hatched lobster larvae, about ⅓ inch long, swim in the plankton layers for up to a month, molting several times. At the fourth molt they become post-larvae and take up the night-prowling, bottom-dwelling life of an adult lobster. Lobsters feed voraciously on crabs, smaller lobsters, sea urchins, fish, and any other animal food that cannot escape. If they escape man and other predators, lobsters may live fifty to one hundred years and reach tremendous size. The largest lobster on record weighed 44½ pounds.

Most lobstering states of the United States have a minimum size limit of 3-3/16 inches from eye socket to end of carapace. In the wild, it takes a lobster from five to eight years to reach this size (about one pound) which corresponds roughly to the beginning of reproductive maturity. However, it is estimated that 90 percent or more of lobsters that reach this age are harvested before they have a chance to reproduce. Berried females are required by law to be released, but the law may not always be observed. If the minimum size were increased to 3½ inches, each female would have had the chance to reproduce at least once.

Attempts at lobster culture in the United States began as long ago as the late nineteenth century with state and federal hatcheries raising larvae for release. For brood stock they depended on berried females purchased from commercial lobstermen. Most of the American programs and their Canadian analogues were discontinued because of high larval mortality and the fact that few of the released juveniles seemed to survive to market size. However, in 1948 the state of Massachusetts reinstated its lobster hatchery on the island of Martha's Vineyard, and most of the fundamental work on lobster culture is due to the scientists there, particularly Mr. John Hughes, the grand old man of lobster culture.

One such advance was in spawning lobsters in captivity. There was never any difficulty in getting lobsters to mate in the lab, but the females almost always jettisoned their eggs without fertilizing them. In the lab, lobsters are usually kept in six inches of water for convenience in handling them and keeping the tanks clean. Hughes discovered that by simply raising the water level to eighteen inches the females would keep their eggs, thus making it possible to contain the whole life cycle of the lobster in the lab. This also makes possible selective breeding programs for such traits as nonaggressiveness, disease resistance, fast growth, and greater proportion of meat to body weight. It was an intuitive discovery, says Hughes, who tried to think like a lobster and realized that if he were in an amatory mood he'd like to be covered up. Eighteen inches happened to be the most economical measure for cutting up plywood to make tanks.

Hughes also has a breeding program for lobsters of pure colors: red, blue, yellow, and albino (normal lobster coloration is a dark, mottled combination). This is useful as a substitute for tagging in growth and survival studies of stocking programs.

Another Martha's Vineyard credit is a circular plastic tank for rearing the newly hatched larvae. Water circulates throughout the tank and keeps the larvae drifting in midwater. If they were to settle on the bottom, the more aggressive ones would eat up the rest. This simple technique saves up to 40 percent of the larvae.

Cannibalism is the most serious problem in lobster culture. To prevent it, it is necessary to place the lobsters in individual cages where they cannot get at each other. This is space-consuming and costly, but necessary. It has the added advantage of preventing the more aggressive lobsters from monopolizing the food and growing fat and strong at the expense of their companions.

A number of basic cage plans have been developed. One places the cages in trays, one deep. This makes for easy feeding and access to the lobsters, but demands the most space of any design. The trays are constantly flushed by seawater, and the shallow depth aids oxygenation. For economy, the trays may be stacked in racks, one above the other. In some plans, automated feeding nozzles would spit food pellets into the trays at prescribed intervals, but this is not yet a reality.

Another system is the care-o-cell, a single layer of cages arranged in a circle. The cages rotate around a central pivot arm. As they do so, they pass beneath spray nozzles that deliver oxygenated water. For commercial-scale production, feeding nozzles could be added.

However, for space economy, the most likely candidate is a multi-storied cage that sits in a deep tank. A gantry crane lifts the cage up for feeding, cleaning, and harvesting. Plans have also been designed for raising caged lobsters on the seafloor, which would avoid the expenses of tanks, plumbing, pumps, and oxygenators.

Diet, of course, is another problem. Originally finely ground beef liver was used to feed the larvae. This was replaced by ground clams or mussels and brine shrimp, which is much cheaper and gives a better survival rate. Older lobsters are fed on chopped fish or shellfish meat. This is still prohibitively expensive on a large scale, but the diet barrier was broken by a University of California (Davis) scientist named Douglas Conklin,* who developed a cheap synthetic ration based on soybeans and milk protein. This produces a lobster of excellent flavor and consistency, but almost snow-white. For the traditional lobster red, Conklin recommends a dash of paprika in the diet. Other researchers have found that cod-liver oil makes any synthetic ration palatable to lobsters, which opens the way to other possibilities.

Both at Martha's Vineyard and in California, researchers have been able to rear lobsters to market size in two years by raising them in water heated to 72° F. The heat source, once again, is power-plant effluent.

Conklin's team has also learned to shorten the long period between laying and hatching. This is done by manipulating water temperatures and daylight hours to make the lobsters think they have passed through two years of seasonal changes. Even so, commercial lobster culture will probably not be a reality before the late 1980's.

Outside the United States, most of the research on lobster culture is being done in France, Britain, and Japan. The French are stressing a hatchery program to replenish wild stocks; in conjunction with this program they have instituted a lobster sanctuary where lobster fishing is forbidden. Juveniles released here will, it is hoped, migrate to other favorable habitats where they will be fair game for lobstermen. Britain has a program similar to that of the state of Massachusetts, but has gone on to construct a pilot grow-out facility that is meant to serve as a model for commercial operations. The Japanese have concentrated on culturing lobsters in floating cages in the sea. They have found that lobsters in these cages need no feeding. They survive and grow by browsing on the fouling organisms that

*Conklin was a student of a scientist who studied under Hughes. This makes him a kind of scientific grandson of the Massachusetts savant.

grow on their cages, plus the larger animals of the plankton. Japanese scientists have raised at least one lobster to the respectable size of four pounds by this method.

Most of the lobster tails sold in restaurants and supermarkets come from one of the many species of spiny lobster. These lobsters, which lack the big crushing claws of the American and European lobster, belong to the Palinurid family, a different family from that of the *Homarus* species. Most of the commercial species belong to the genus *Panulirus*. (No, this switch of spelling is not a typo. It is probably some taxonomist's outburst of playfulness.)

Most species of spiny lobsters inhabit warm, semitropical or tropical waters, although the renowned European *langouste* ranges from the Mediterranean north to Britain.

From the culturist's viewpoint the spiny lobster has many virtues: good food conversion efficiency, disease resistance, a short incubation period (as little as three weeks in some species), and a low rate of cannibalism. On the other hand, it has a very complex larval history, beginning as a transparent, long-legged, spidery creature called a *phyllosoma* (Greek for "leaf body"). After many molts, the phyllosoma becomes a *puerulus* (Latin for "little boy"), a trifle under an inch long. This is equivalent to the post-larval stages in shrimps and true lobsters. Due to lack of knowledge of the larval dietary requirements, no one has yet been able to rear a spiny lobster to the puerulus stage.

Crabs are of considerable commerical importance, but so far the only one to be cultured is the Oriental species *Scylla serrata*, which is raised for its roe in Taiwan and Southeast Asia. Most often the young crabs enter fish ponds accidentally and just grow. However, even with no management at all they produce a considerable yield. In Taiwan, just-mated females are sometimes cultured in small, masonry-lined ponds and fed on snails, trash fish, and other cheap animal foods.

A small-scale cultural project is under way in Japan, where fisheries scientists take berried females and raise the larvae to juvenile size. The juveniles are sold to fishermen's cooperatives for restocking.

CHAPTER 10

The Succulent Oyster and Other Mollusks

Man has eaten oysters and clams sinced time immemorial, as great shell heaps at prehistoric seaside campsites testify. Long regarded as a staple food for the poor, being plentiful, easy to harvest, and therefore cheap, they are now in the delicacy category in most countries. However, with a little help from man, these simple, rather primitive animals and their relatives promise to outstrip finfish as a source of protein. Not only do they reproduce abundantly, they eat very low on the food chain and have an excellent food conversion efficiency, often approaching 1 to 1.

Oysters and clams belong to the large invertebrate phylum of mollusks, a name derived from the Latin word for "soft" and referring to their soft, boneless bodies. More specifically, they belong to the class of *Pelecypoda* (Greek for "hatchet feet"), along with mussels and scallops. They are more commonly known as bivalves, from the two halves of their hinged shells. *Valve*, by the way, comes from a Latin word for the leaf of a door. Other important classes of mollusks include the gastropods (snails and slugs) and the cephalopods (squid, octopus, and nautilus). Members of all these groups are eaten by man, but the bivalves are by far the most important.

Bivalves evolved as basically sedentary bottom-dwellers (some species attach themselves to vertical surfaces such as rocks and pilings). They are filter feeders, living on microscopic plankton organisms, chiefly one-celled algae that they strain from the seawater and trap in the mucus of their gills. A single oyster may pump as much as one hundred gallons a day of water through its system, extracting food and oxygen. Lips at the beginning of the digestive tract screen out too-large particles, most animal plankton, and other unsuitable matter. The bivalve gets rid of this by periodically opening and rapidly closing its shell; the resulting waste product is called pseudofeces.

Cross-section of an American oyster, Crassostrea virginica. *The mantle is the organ that secretes the shell; the muscle shown is the adductor, which closes the shell firmly; the palps are the lips at the opening of the digestive tract that filter out unsuitable particles. Other terms in the diagram are self-explanatory.*

COURTESY OF NATIONAL MARINE FISHERIES SERVICE, MILFORD LABORATORY

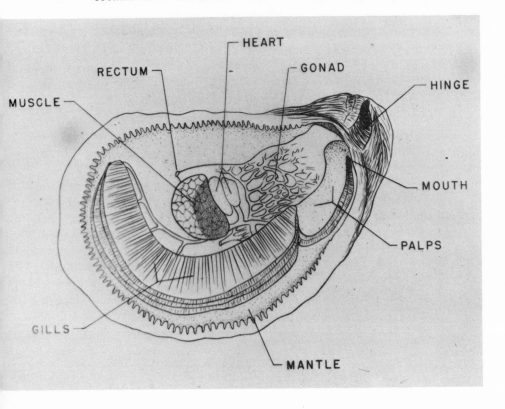

There are many species of oysters, but only a few are cultured. The most important commercially are the Japanese oyster (*Crassostrea gigas*), the flat European oyster (*Ostrea edulis*), the Portuguese oyster (*C. angulata*), the American oyster (*C. virginica*), which is native to the Atlantic coast, and the slipper oyster (*C. eradelie*), which is the chief oyster of the Philippines. The pearl oyster is actually not an oyster but a large mussel. Although pearls are cultured on a large scale in Japan, pearl culture is not generally considered a part of aquaculture, perhaps because pearls are not eaten.

Although the different species vary in their schedules, oysters spawn when the water temperature rises above a certain critical level. Some tropical species spawn around the year; in temperate regions the spawning season coincides roughly with the summer months. The flesh of oysters after spawning is lean and watery, hence the old adage that oysters should not be eaten during the months without an *R*. Another reason is probably the danger of spoilage in hot weather, common in pre-refrigeration days.*

The American oyster, triggered by rising water temperatures, spawns profusely. Females release millions of microscopically small eggs, about 1/600 inch in diameter. A good-sized female can produce on the order of 100 million eggs in a season. The males, nearby in the oyster bed, simultaneously release milky clouds of sperm. Chemicals in the sperm and eggs each stimulate oysters of the opposite sex to release their own products, which synchronizes the process. Fertilization is almost instaneous, and always takes place within half an hour.

The fertilized eggs develop very rapidly, and in about thirty hours they become straight-hinge, or *veliger*, larvae, complete with a tiny shell that resembles a hardshell clam's and measures about 1/400 inch across. The larvae also have a wheel-like organ, the velum,

---

*Many oyster specialists believe that the "no R month" rule was derived from the European oyster, which broods its larvae in its mantle chamber. The developing larvae give the oyster a gray tone when they are small; when they are almost large enough to emerge and set, they give the parent oyster's flesh a black appearance. Though harmless, these unappetizing color phases are known as gray sickness and black sickness. In addition, the tiny shells of the larvae give the oyster a gritty, unpleasant consistency. However, none of this applies to the American oyster, which does not incubate its offspring within its shell. American oysters are usually in prime, plump condition in May and June.

equipped with cilia, with which they can swim for short distances, mainly up and down in the water column, seeking food among the other plankton organisms. Tidal currents transport the larvae to new locations, often far from the bed where they were spawned.

They lead a drifting, planktonic life for about two weeks until they are ready to settle down. During this time 90 percent or more of the larvae are devoured by fish, shrimps, barnacles, sea squirts, and a host of other predators, or die of natural causes such as drifting into water of the wrong salinity or oxygen content.

Week-old oyster larvae, highly magnified, resemble tiny clams. The velum, a wheel-like organ with which the oyster larva kicks itself through the water, is visible on the larva at right center.

COURTESY OF LONG ISLAND OYSTER FARMS

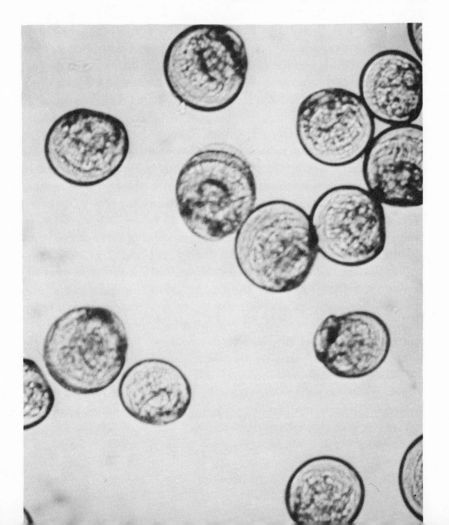

The surviving larvae develop a fleshy foot around their ninth day, and about the same time they develop a rudimentary eyespot that can probably distinguish light and dark.

After about two weeks comes the most perilous moment of an oyster's life: the transformation from a planktonic organism to a fixed bottom-dweller, otherwise known as setting. The larvae sink to the bottom and crawl about seeking a hard, bare surface to attach themselves to. If they do not find one, they can push off from the bottom and swim away to try their chances elsewhere. The search may continue for two or three days, but then the larvae must settle down for good. If they settle on sand or mud, they die. Those that are lucky find a rock, a shell, a discarded tire, tin can, or other hard object. They secrete a dab of glue with their byssal glands, then immediately turn on their left sides and cement themselves fast near the hinge end of their shells. From this moment on they have lost the ability to move themselves. Velum, foot, and eyespot disappear in a strenuous metamorphosis that many oysters do not survive. It is estimated that only one percent of the larvae that have escaped the hazards of free-swimming life—one in one thousand of all the new-hatched larvae—survive the involuntary gamble of setting.

The spawning story is pretty much the same for all the *Crassostrea* oyster species. In the case of the flat European oyster, however, the eggs are fertilized inside the female, which inhales sperm-laden water, remain inside their mothers for about eight days, emerging as *pediveligers* (larvae equipped with velum and foot).

Under favorable growing conditions, an oyster can reach sexual maturity in less than a year. A peculiarity of oysters is that they change sex frequently. A majority of oysters begin as males, changing to females as they grow older. However, females often become males after they discharge their eggs. After the spawning season ends, both sexes resorb their sex organs and spend the winter as semi-neuters. When the next spawning cycle begins, they blossom out again, unpredictably, as male or female.

In the wild, it takes oysters from four to five years to reach half-shell size, although in unusually rich feeding grounds or warm water they may reach that size in as little as two years. Legal minimums vary from state to state and country to country, but the average is probably two to three inches.

The newly set larvae, or spat, are vulnerable to many dangers. Storms may cover them with silt and suffocate them. Starfish, oyster drills (a kind of snail), and rays prey on them. Crabs eat them when they are small. Mussels and other fouling organisms, including other oysters, grow on them, sometimes suffocating them and always competing with them for food and oxygen. Seaweed can foul them, too. Unable to move, the oysters' only chance of survival as a species lies in numbers.

Disease can be disastrous to the oyster population. On the east coast of the United States a protozoan disease called MSX or *Minchinia* wiped out the once-flourishing oyster industry of Delaware Bay and many of the most productive oyster grounds of Chesapeake Bay. A small percentage of the oysters were resistant to the pest and presumably passed this resistance along to their progeny, but it takes decades to regenerate, from such a small start, an oyster population that once numbered in the billions. A fungus disease, *Dermocystidium*, also wreaked havoc on the East Coast. In Europe and Dutch shell disease, which attacks the shells of oysters and renders them fragile and crumbly, ravaged the oyster beds in the 1940's and is still a major hazard.

Oyster culture probably began with the Romans, albeit on a rather limited scale. Near the end of the second century B.C., a businessman named Sergius Orata was cultivating oysters on a large scale in Lake Lucrinus, near Naples. The source of our information, the famed Roman historian Pliny, does not say whether Orata brought in seed oysters from elsewhere to fatten them up in the brackish lake or whether he simply broke up the naturally occurring clumps of oysters and spread them out for better feeding and breathing. It is recorded that Orata filled the lake with his oyster enclosures, so that the other wealthy men who owned villas on the shore brought suit against him. Following Orata, other Romans learned that they could collect oyster larvae by placing out stakes in the water for them to set on. This practice came to an end with the crumbling of the Roman Empire, and almost nothing was done in the way of oyster culture in Europe until it was revived by a French biologist named Coste in the mid-nineteenth century.

Coste, backed by the Emperor Napoleon III of France, hoped to develop large-scale culture of oysters and finfish to provide cheap, tasty, nutritious food for everyone. His plans collapsed when Napoleon lost his throne in the Franco-Prussian War of 1870. But he did accomplish one thing. By studying Italian oyster culture at Lake Fusaro, near Naples, he learned about collecting spat on objects hung in the water. The oystermen of France, faced with nearly exhausted beds, tried all kinds of collectors. Coste made his study in 1853. It took over a decade of trial and error before a French culturist hit on the idea of hanging ordinary roof tiles dipped in lime in the water for the spat to set on. This simple technique saved the dying French oyster industry.

Why this worked was not understood at first. Later studies found that oyster spat, at least of the flat European oyster, prefer clean surfaces for setting, and they also like lime, perhaps because of a chemical similarity to oyster shells. At any rate, they set bountifully, sometimes as many as three thousand to a tile. The oyster growers were also delighted to find that the spat could be easily cracked loose from the brittle lime coating when it came time to transplant them for fattening, whereas if they set on stones, bricks, or old oyster shells they often broke in the process of prying them free. The last refinement was stacking the tiles in pairs, crisscrossed, and six pairs high. this made for a more stable structure, not so easily knocked over by the current.

Ceramic tiles are still used for collecting spat in France, although some oyster growers are shifting to plastic tiles, which are much lighter and easier to handle. The spat are cracked loose and transplanted to growing beds when they are about thumbnail size.

The center of French seed oyster production is the Gulf of Morbihan, on the south side of the peninsula of Brittany. Seed oysters from here are shipped out to other locations in France and elsewhere in Europe to be fattened. French oysters are carefully tended. The oyster farmers turn them over with rakes and pitchforks to make them grow evenly. They are moved several times, from shallow water to deep and back again. The average flat oyster in France takes four and a half years to reach market size and is moved three times in that period.

A certain kind of diatom in the water has the peculiar property of turning the oyster's gills green, a condition much prized by demanding French gourmets. To gratify this taste, some oysters are given a final half year in shallow ponds called *claires*, which are dug in the salt marshes that fringe many French shorelines.

Although the French have developed the raising of oysters for the table to a fine art, they have done little to go beyond the dependence on wild spat for their seed oysters. Not much work is being done with hatcheries anywhere in Europe, for that matter.

Oyster culture of a sort dates back to the 1850's in the United States. Oysters dredged from the rich grounds of Delaware and Chesapeake Bays were transplanted to the depleted beds of northern New Jersey and Staten Island for a summer's fattening. (Experts would point out that "fattening" is not quite the right word, since oysters contain little fat. What makes them plump and rich is glycogen, a form of starch secreted by animals as a way of storing energy.)

By the 1880's overexploitation had reached the danger point, and farsighted observers were beginning to urge that something be done to propagate the declining oyster.

Indeed, as early as 1879 a scientist at Johns Hopkins University in Maryland had succeeded in hatching oyster eggs in the lab. It was a long time, however, until marine biologists were able to raise oyster larvae to the point of setting—1920, in fact. The next step was to spawn the oysters artificially, achieved in the 1940's by Victor Loosanoff and his associate H. C. Davis at the U.S. Bureau of Commercial Fisheries lab in Milford, Conn. Loosanoff combined artificial spawning with the culture techniques perfected by W. F. Wells in 1920. Oyster hatcheries today use one or both of these systems, modifying them on the basis of constant research.

It is a simple matter to make oysters spawn. Basically, all you have to do is trick them into thinking it is spawning time (or, more precisely, trigger their physiological responses) by raising their water temperature to 77° F. and keeping it there for a few hours. In practice, the oysters are first held for several weeks at 50° F. to get them in proper starting condition, with all their old eggs and sperm resorbed. The water temperature in their tanks is then gradually raised to 73° and held there for five weeks, after which the oysters are cooled down to 66° to prevent any of them from spawning prematurely. Finally, the temperature is raised to 77°, and the oysters respond by discharging sperm and eggs within an hour.

Brood oysters are selected for size, shape, and growth rate. Since brood oysters are usually taken from the growing beds, they don't come equipped with lab records of their growth. Fortunately for culturists, the shell itself supplies clues. Large growth rings indicate a fast growth rate, and so does a relatively thin shell. Shape is important for the half-shell trade. Most oyster lovers prefer a smooth, regular shell. (For the shucked-oyster trade, which includes canned and frozen oysters as well as fresh oysters, shell shape doesn't matter because no one sees the shell.)

Size is a clue to sex: oysters 1½ to 2½ inches long are likely to be males; above 3 inches they are pretty certain to be females. There should be about three males to every two females in the spawning trays to ensure that all the eggs get fertilized.

Once spawning is completed, the water containing the clouds of fertilized eggs is pumped into big plastic rearing tanks with conical bottoms. The conical bottoms collect wastes and make cleaning the tanks easier. Twice a week the tanks are emptied and the larvae are screened to grade them for size. Undersized larvae, which are about 80 percent of the total, are discarded. The remaining larvae are destined to be fast growers.

Diet is critical for the larvae. There are several methods for supplying them with the necessary algae. One, the Wells-Glancy method, uses the algae naturally present in the local seawater. The water is centrifuged to get rid of animal plankton and the larger algae, since oyster larvae clan ingest only very small cells. The remaining algae are then allowed to multiply in a tank for twenty-four hours. Algae-bearing water is pumped out periodically and passed through the larval rearing tanks. It is replaced with filtered seawater, which gives the algae the nutrients they need to build up their numbers again. If not enough algae are present naturally after cent-rifuging, more are added from an algae culture. Some hatcheries, however, prefer to use their own pure algae cultures exclusively.

There are thousands of species of one-celled algae in the sea, but not all of them are equally good as oyster food. Some of the species most popular with oyster scientists are *Isochryisis galbana, Thalas-siosira pseudonana,* and *Skeletonema costata.* The oysters are fed on cultures of these or other suitbale species from hatching until they are big enough to plant out.

Top: At commercial hatcheries, oyster larvae are raised in tanks until ready to set out. Water quality and temperature are closely controlled, and the oysters are fed on a diet of selected species of algae.

COURTESY OF LONG ISLAND OYSTER FARMS

Bottom: Oyster larvae are screened periodically to grade them for size. Undersized larvae are discarded. There are about one million oyster larvae on this screen.

COURTESY OF NATIONAL MARINE FISHERIES SERVICE, MILFORD LABORATORY

Experience has shown that no one species of algae is a completely balanced oyster ration. However, scientists at the University of Delaware have discovered that a pure diet of *T. pseudonana* speeds the sexual maturity of oysters while stunting their growth. While this would be a definite detriment for oysters destined for the market, it has advantages for the producer of seed stock, since he can accommodate more brood animals in less space and on a smaller food supply per oyster.

Although algae grow with luxuriant abandon in the sea, in the lab they can be a delicate proposition. They are very particular as to salinity, water temperature, nutrients, and light, and each species has a different preference. This means that for the most efficient algae production, each species you grow must be raised separately under its own optimum conditions.

Even so, disasters can and do happen. An algae culture can "crash" when the number of cells get out of balance with the nutrient supply. A power failure may chill the busily dividing little cells into inactivity. A leaky roof can alter salinity in the algae vat beneath it to a fatal degree, or contaminate the culture with algae-devouring protozoa.

Raising algae is an expensive business even if no accidents occur, for the cultures need daily care. It is estimated that about one-third of the cost of running an oyster hatchery is due to the expense of raising algae. It is no wonder that one of the long-term goals of oyster scientists is developing a synthetic oyster chow that can be made cheaply from vegetable products.

With the oyster larvae's diet disposed of, let us return to the larvae. In the hatchery they are ready to set in about nine days. At this point they need something to set on. Traditional oystermen, like those of Chesapeake Bay, strew the bottom with old oyster shells for the spat to set on. These are called "cultch." In the hatchery, the cultch is apt to be small fragments of shell, big enough for an oyster larva to attach itself to at the critical moment of metamorphosis but small enough not to interfere with the growth of the baby oyster's shell. In effect, the hatchery oyster is cultch-free, which means that it has an excellent chance of developing a symmetrical shell, much in demand for the half-shell trade. Also, it will not need to be knocked free from

Oyster hatcheries maintain an impressive array of algae seed stocks to provide food for the oyster larvae.

COURTESY OF LONG ISLAND OYSTER FARMS

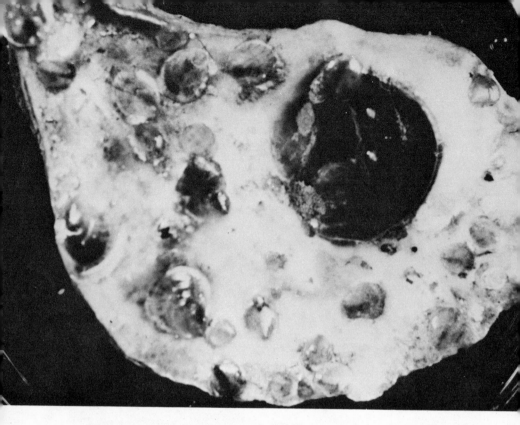

Young oyster spat have set on an oyster shell, their favorite choice of attachment. In a pinch, any hard surface will do.

COURTESY OF NATIONAL MARINE FISHERIES SERVICE, MILFORD LABORATORY

clumps of oysters, rocks, or whatever else it is attached to when it is harvested, since the original tiny fragment of cultch is all it needs.

Scientists at the Virginia Institute of Marine Science (VIMS), which has had great success in breeding oysters, recommend hanging sheets of Mylar plastic in the setting tanks. The spat set readily on the flexible plastic and can be easily removed by pulling the plastic over a sharp edge. As it bends, the oysterlings pop off and fall into a collecting basket. At the University of Delaware, the oyster experts get excellent results by setting their spat on mats of Astroturf, another soft, flexible plastic.

Fifty to seventy of every one hundred larvae that set live through the process of metamorphosis. The remaining spat should be thinned out to keep them from becoming too crowded as they grow. The VIMS hatchery manual recommends fifteen to twenty set oysters per square inch of collector surface. Once safely past metamorphosis, the oysters are switched from cultured algae to straight, raw seawater. This cuts the cost of feeding them, but brings the danger of introducing parasitic protozoans and the eggs and larvae of fouling organisms such as sea squirts. These pests can be controlled by chemical treatment and by filtering the water. It is also essential to keep the raceways free of waste by keeping a constant flow of water through them. If this is not done, the smaller oysters may suffocate in the feces and pseudofeces.

The spat are kept on their collectors until they measure about ⅜ inch across. They are then removed and transferred to raceways or big trays through which water flows slowly. Here they are kept until they are large enough to plant out.

Technically, the oysters could be planted out as soon as they have completed setting. In practice, however, their chances of survival are better the larger they are when planted out.

We must digress here a moment to explain that certain states, such as Maryland, consider the sea bottom to be public land, to be enjoyed equally by all. In principle, this is beautifully democratic and fair. In practice, it results in rundown, overexploited oyster beds. It simply does not pay for oysterman A to restock an oyster bed and combat predators when his fellow oystermen B, C, D, and so on all the way down the alphabet then help themselves freely and legally to his oysters. So everyone grabs what he can and leaves it to nature and the state to replenish the beds, to the eventual detriment of all.

Other states, such as Virginia and New York, permit shellfish growers to own or lease plots of the seafloor for their private use. No outsider is permitted to take shellfish from there any more than he would be allowed to help himself to vegetables from a commercial farm. As a result, it is profitable for the growers to take care of their beds, and production flourishes year after year. The same principle holds true in Europe and Japan.

A number of hatcheries simply supply seed oysters to growers. Others combine the hatchery with grow-out beds. One such hatchery is the Long Island Oyster Farm. The LIOF keeps its oysters indoors until they are about ⅜ inch in diameter, which is roughly the breadth of an adult human's little fingernail. Then they are placed in cages with stacks of wire-screen trays and hung in the cooling lagoon of the Long Island Lighting Company's power plant at Northport. The lagoon, heated by the discharge from the power plant, stays at a pretty steady 20° F. above the ambient water temperature in neighboring Long Island Sound, a few hundred yards away. The same water, filtered and treated, is used in the hatchery. It is a cozy arrangement. The power plant stands on one bank of the long, narrow lagoon, the hatchery on the facing bank.

The oyster cages are hung between sturdy piers. Each one holds about 25,000 baby oysters. The constantly warm water accelerates the oysters' growth rate so that they reach 1½ inches in about two months. They are then planted out on the several thousand acres of bottom that the LIOF controls in eastern Long Island Sound, where the water is clean. There they reach market size in two and a half years, rather than the four years it takes naturally.

A staff biologist cleans and inspects juvenile oysters at the Long Island Oyster Farms' hatchery at Northport, Long Island. The oysters will be returned to the heated waters of the lagoon beside the hatchery, where they grow rapidly for eight to ten weeks.

COURTESY OF LONG ISLAND OYSTER FARMS

These young oysters at the Long Island Oyster Farms are ready to be removed from the heated lagoon and transferred to beds in Long Island Sound. The oysters are approximately four months old; at the tips of some of the shells can be seen some of the tiny cultch fragments to which they cemented themselves when they set.

COURTESY OF LONG ISLAND OYSTER FARMS

The LIOF hatchery is not big enough to meet the market demand; so a good proportion of spat are collected near New Haven. However, an expansion of the hatchery is in the works. Some spat is trucked down from a branch hatchery in Maine.

The oyster beds are patrolled periodically by scuba divers who check them for the oysters' worst natural enemies: starfish and oyster drills.

Not a fish at all, but a member of a primitive phylum of creatures called echinoderms (Greek for "spiny skin"), the starfish creeps along the bottom on its five sucker-lined arms in search of oysters, clams, and other bivalves. Thanks to its suckers, it can climb rocks and pilings, but it cannot swim. Blind and brainless as a starfish is, it has sensors that tell it when prey is near. The starfish attaches its arms to the two shells of the hapless bivalve and pulls on them. Eventually the oyster weakens and lets its shell be opened. The

149

starfish everts its stomach inside the shell and digests the oyster. A numerous starfish population can wreak havoc on an oyster bed.

The oyster drill is a small snail, usually an inch or less in length. It bores a hole through the oyster's shell with its file-like tongue, aided by a secretion that loosens the natural cement that holds the limy crystals of an oyster's shell together. Then it inserts its tongue and rasps away the living oyster's flesh. Despite its superb boring equipment, it takes an oyster drill a long time to work its way through a thick shell. Therefore drills prey mostly on spat and young oysters, and they can wipe out a year's crop of spat in a short time.

Starfish can be killed by dumping quicklime over the oyster beds. The lime attacks the starfish and causes them to disintegrate within twenty-four hours. The oysters close their shells and are not harmed. Another much-used method is the "mop," a long iron beam to which lengths of light chain are attached. Each chain has at its end a bundle of rope yarn about six feet long. As a boat drags the mop over the oyster bed, starfish become entangled in the yarn. The mop is hoisted aboard at intervals and dunked in boiling water to kill the starfish.

This is what boring sponges can do to living oysters' shells.
COURTESY OF NATIONAL MARINE FISHERIES SERVICE, MILFORD LABORATORY

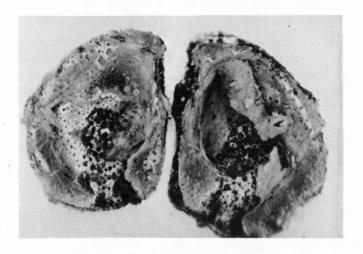

Top: Starfish are among the worst natural enemies of oysters.

Bottom: The starfish mop is an effective weapon against the primitive but destructive echinoderms. Starfish-laden rope yarns are dipped in boiling water to kill the entangled starfish.

Some old-time fishermen, ignorant of starfish biology, used to cut the starfish into pieces and throw them back in the sea. Instead of killing the starfish, this increased their numbers, for every piece regenerated itself into a complete new starfish.

One was of controlling oyster drills is to use a seagoing plow to turn over the seabed and bury them. However, this must be done before planning out the oysters; otherwise the oysters are plowed under and killed too. Another technique is to vacuum the bottom with a suction dredge, which also gets rid of starfish. This, too, must be done before setting out the young oysters.

An alternative method, useful on a small scale, is to dredge up the oysters and dip them in highly concentrated brine. The oysters are then exposed to the air for several hours. This treatment usually gets rid of predatory flatworms, starfish, and many fouling organisms as well as the drills.

Among these creatures are boring sponges and mud-blister worms, which are detrimental to the oyster even though not usually fatal. They bore into the oyster's shell and make a home there. This weakens the shell and ruins it for the half-shell trade because of its unattractive, half-rotted appearance. It also slows the growth of the afflicted oyster, which consumes a great deal of energy replacing the shell these organisms have destroyed.

The Long Island Oyster Farm has its own packing plant at Greenport, near the eastern end of Long Island. Here the dredge boats unload large steel baskets that hold fifty bushels of oysters. The dredge baskets are dumped onto a moving belt, and hoses blast off the mud that clings to the oysters. Workers pick out and discard crabs, worms, and other unwanted marine life that comes up with the oysters. (No machine can do this task.) Other workers quickly sort the oysters by size and discard those wtih defects. Those oysters that are destined to be eaten on the half-shell are packed in shipping cartons, color-coded for oyster size, and rushed to a chilling room.

Oysters are culled, sorted, and cleaned on the moving belt at the Long Island Oyster Farms' packing plant.

COURTESY OF LONG ISLAND OYSTER FARMS

Refrigerated trucks take them to wholesale dealers.* Surprisingly, the LIOF and other eastern growers ship oysters by the truckload to the West Coast, which has plenty of oysters of its own.

The oysters to be shucked (the proportion varies according to the demands of the market) go to another conveyor line, where they are dumped in small piles before workers who deftly open them by hand with a special knife and dump the meat into a pail. No one has yet devised a successful machine for shucking oysters, through many people have tried. A skilled shucker can shuck up to six hundred oysters on hour.

The shucked oyster meats are washed and packed in sterile, one-gallon cans, under very hygienic conditions. A state inspector checks the plant periodically to make sure that everything stays sanitary.

Although the LIOF takes advantage of technology to speed the growth of their oysters, their basic philosophy, like that of most growers, is to leave as much to nature as possible. Working from the opposite viewpoint are a group of scientists at the University of Delaware's College of Marine Studies. These men have been working for years to develop a completely self-contained, closed-cycle system for rearing oysters all the way to market size. The method works on an experimental scale, although it is not yet commercially practicable. If and when perfected, it will enable oysters to be grown anywhere, even in the middle of Nebraska.

The advantages of such a self-contained system are many. You can control every aspect of the oyster's environment: water temperature, oxygen, and nutrition. For example, the Delaware group have been breeding improved algae for a number of years and still use cultures initiated in the 1950's. You have no problem with predators, parasites, or pathogens. You can conveniently run a selective breeding program, since your breeding stock is right at hand. And, by adjusting the mineral content of the water, you can even control the thickness of the oyster's shell.

On the other hand, you also have the expenses of keeping up water quality and disposing of the oysters' wastes, as well as the other usual intensive aquaculture problems.

*Most LIOF oyster end up in restaurants rather than in retail fish markets.

The Delaware scientists have developed a prototype oyster green-house, where they raise algae in tall, tubular fiberglass tanks. The silolike shape permits maximum algae production per square foot of ground. The tanks are outside the greenhouse; the oysters are inside in raceway trays, where they reach maturity in one-fourth the time they need in the wild.

So far this plan seems to be unique in the United States, except for that of one short-lived firm that built an oyster greenhouse in Lamar, Colorado, using the effluent of a power plant for heat and an artificially formulated seawater.

The LIOF and a number of other growers raise modest tonnages of the European oyster along the coast of Maine. Although fisheries officials and environmentalists are justifiably very cautious about introducing an alien species of animal, there is apparently no danger of the European oyster's crowding out American oysters in Maine. The Europeans have been planted only where there are no native oyster beds, and the prevailing currents may prevent their larvae from spreading.

On the West Coast, the native Pacific or Japanese oyster, *Crassostrea gigas*, is cultivated on a large scale. *Gigas* is usually sold shucked, due to local preferences. The centers of oyster culture are protected waters such as Washington State's Puget Sound, where the Olympic oyster, *O. lurida*, is also grown. Conditions favor spat collection only at two locations: Dabob Bay in Washington and Pendrell Sound in British Columbia. Several hatcheries have been established to supply growers with seed oysters.

The Japanese have been raising oysters since 1673, when a clam grower accidentally found that oyster spat would set on bamboo stakes driven into the bottom. Today, Japan has one of the most advanced systems of oyster culture in the world. Instead of the old-fashioned and inefficient method of growing oysters on the sea-bed, the Japanese have specialized in growing them off the bottom, suspended from rafts or cables buoyed up by floats. The seed oysters, collected on scallop or oyster shells, are strung on long wires that hang down from the raft or cable. The wires range from twenty-four to 50 feet long, depending on the depth of the water, and the cultch shells are spaced out about eight inches apart to permit free water circulation. The wires stop short of the bottom.

In Japan's Inland Sea, a worker winches up a heavy string of oysters to harvest them. Many more such strings hang beneath the sturdy bamboo raft.

This off-bottom culture offers several advantages. First of all, the oyster grower is using practically all the water column, instead of just the thin layer at the bottom. Second, the oysters are better exposed to food and oxygen-bearing water currents, which spurs their growth rate. Third, by keeping the oysters several feet off the bottom, it keeps them out of reach of starfish, oyster drills, and other nonswimming predators.

Off-bottom culture is not practiced in the United States and Canada because it conflicts with other uses of the water. You cannot go sailing, powerboating, water-skiing, or even fishing if the water is covered with strings of rafts or anchored cables. And few people who own a shorefront cottage would welcome a vista of oyster rafts. However, it has been done commercially where recreational use of the water is not important. United States fisheries scientists as long ago as 1956 raised oysters on rafts in a small estuary on Cape Cod. The oysters grew to market size in two and a half years, while oysters on the local beds took four to five. The raft-grown oysters also had shells of more uniform shape than those that grew on the bottom. In the Inland Sea of Japan, that country's major oyster-producing area, raft-grown oysters reach minimum market size in less than a year. Of course, this record is helped by the fact that the Japanese market size is smaller than that in Western nations.

As in other parts of the world, Japan has many areas where oysters grow well but do not reproduce. Most of the spat is collected in a limited area around Hiroshima and grown at various locations around the Inland Sea. There is another seed oyster center in northeastern Japan, which produces about 2 billion spat a year. About half of these are exported to the United States, Canada, Australia, and several countries in western Europe, including even gastronomically haughty France. For long-distance shipment, the seed oysters are hardened by stringing them on racks where the tide will leave them exposed to the air for four to five hours twice a day.

The other major Asian oyster-growing nation is the Philippines, where oysters are as much a staple food as fish-and-chips in England. The oysters all come from wild spat, of which there are several species in Philippine waters, but the most important species is the slipper oyster (*Crassostrea eradelie*), which is the biggest and most abundant.

Filipino oyster farmers collect their spat on long bamboo poles driven into the bottom in the intertidal zone and leave them there to grow. Government scientists are trying to promote a more efficient system that uses Japanese-style strings hung from bamboo frameworks. (Bamboo is cheap and plentiful in the Philippines.)

With either method, the oysters grow very rapidly in the warm, food-rich coastal waters of the Philippines. They reach minimum market size (three inches) in only six to nine months—a rate that is the envy of American and European oyster farmers.

The typical Filipino oyster farm is a small family or even one-man operation, covering half an acre or less of bottom area. Often the oyster grower has to build a hut on stilts over his oyster plot and live there during the harvest season to thwart poachers.

Worldwide, great progress is being made in every phase of oyster culture, from spawning to harvesting. However, pollution and silting threaten many traditional oyster-growing areas. Silting buries and smothers the oysters; however, it can be countered by off-bottom culture. Pollution is a graver problem; even if the pollutants do not kill the oysters, the oysters may concentrate toxic substances to a point that is dangerous for humans to eat. Against this there is no countermeasure except strictly enforced pollution-control regulations.

Mussels are number two among cultured mollusks. Extremely prolific, hardy, and fast-growing, they produce more protein per acre than any other known organism. Like the other bivalves, the mussel's meat is high in protein—14 percent as compared to 13 percent for beef—low in fat, and rich in vitamins and minerals. Although generally regarded as a nuisance in America, mussels are a delicacy in many countries of Europe and Asia. The famous French dish *moules marinière* (mussels barge-crew style) is one of many ways of preparing this delicious, orange-fleshed mollusck.

The spawning and larval life of mussels is much like that of oysters, with a very important difference. When it comes time for the mussel larva to settle down to a sedentary life, it does not attach itself permanently to the substrate, as the oyster larva does. It anchors itself with a tuft of tough fibers called byssal threads, which it spins with a secretion from its byssal gland. The mussel can slough its byssal threads and crawl to a new location; in fact, it may move several times before finally settling down.

The mussel's first home is usually something filamentous, such as the branches of a hydroid (a semimicroscopic water animal related to the corals and sea anemones), or to certain threadlike seaweeds. Later they move on to harder substrates such as rocks, stones, concrete seawalls, wooden pilings, and other shellfish. They may even attach themselves to broad-bladed seaweeds such as kelp. When small, mussels can be carried considerable distances by tides and currents; some researchers think they create a bubble of gas within their shells to buoy them up.

Mussels like to settle in large colonies called banks or shoals. In the wrong place they can be a considerable nuisance, as when they block the intake pipes of a power station's cooling system or half-smother an oyster bed. Because of their crowding together and the resulting competition for a limited share of food and oxygen, wild mussels grow more slowly than their cultured counterparts.

The country with the longest history of mussel culture is France, and there it dates back to the thirteenth century. According to tradition, an Irish sailor named Patrick Walton was shipwrecked on the coast of France near La Rochelle. Instead of making his way back to Ireland, he decided to stay where he was and make a living by catching the numerous seabirds. Before he could do this, however, he had to have some way of traveling over the muddy tide flats, in which he would have sunk deeply at low tide. From the wreckage of his ship he built a little sledge-boat that would skim over the mud. Next, he fashioned a large net that he strung between tall poles that he drove into the mud.

Walton soon found that his net poles were covered with mussels, and in a few months he saw that they were growing much faster than those clinging to the rocks. Delighted with his serendipitous crop, Walton experimented further and eventually discovered the method that is still standard in France today.

Loosely woven ropes about ten feet long are hung in the water in early summer. Within two weeks, newly set mussels have filled the crevices between the rope yarns. The ropes with their cargo of seed mussels are then wrapped in an open spiral around tall poles that are driven into the mud or sand.* The last foot or so of the pole above the seabed is sheathed with smooth plastic to keep crabs and other predators from climbing up. Although the tides leave the mussels high and dry for several hours each day, during which time they cannot feed, the mussels grow so rapidly that they are two inches long and ready for harvest in a year. Within a few months, in fact, they cover the whole pole and encrust it several layers deep. The outer layers of mussels must be removed to keep them from choking out the inner ones, and to keep them from breaking off their own weight and being lost on the bottom. The surplus mussels are placed in net bags and wrapped around new poles, or else simply slung between the poles.

The major French mussel-producing area is now the northern coast of Brittany, on the English Channel. Here the plankton-rich waters and strong currents are excellent for mussel growing, but the very high tides (twenty feet and up) sweep mussel larvae out to sea before they can set. Therefore, the Breton growers depend on seed mussels collected around La Rochelle and further south.

Mussels are also raised in the Netherlands and in Norway. The Norwegians have a neat trick of wrapping their mussel sets in long bags of expandable plastic mesh. As the mussels grow, they crawl through the mesh and attach themselves to the outside of the bags, which are hung in the water without being wrapped around a support.

But the leading mussel-growing country nowadays is Spain, where mussel culture began in a tentative way about one hundred years ago and became big business after the end of World War II. Spanish mussel culture is concentrated on the northern coast of Spain, where there are deep, sheltered, fordlike bays.

The mussels are grown on rafts, an adaptation of the technique the Japanese use with such success on oysters. Most of the seed comes from wild mussels collected from the rocks of the coast; but about 30 to 40 percent settles directly on special collector ropes made of fuzzy materials. In either case, the seed mussels are wrapped around

---

*The poles run about thirteen feet long; they are sunk into the bottom for about half their length to keep them from toppling over.

grow-out ropes with plastic mesh, through which the young mussels soon crawl and space themselves out. In the best locations, the mussels reach market size in as little as eight months. In poorer locations, such as the inside of a bay that is crowded with other mussel rafts, it may take twelve to fourteen months. The reason for this slower growth is the marine equivalent of overgrazing. The mussels' efficient filtering systems extract about 35 percent of the plankton in the water that passes by them; so that by the time the water reaches the inmost rafts it is picked rather bare.

Even so, a raft of the commonest size, about 60 by 60 feet, holds 600 to 800 ropes and produces anywhere from 66,000 to almost 200,000 pounds of mussels per year. Some of the newer rafts measure 66 by 66 feet and hold 1,000 ropes, with a proportionately greater yield.

As of early 1979, Spain's five mussel-growing bays held about four thousand rafts. Spanish fisheries officials estimate that there is room for another thousand rafts before production reaches its limits.

In the United States, mussels have traditionally suffered from a poor reputation because they grow profusely around sewage outlets (so do oysters). However, if they are grown where the water is unpolluted, they are perfectly clean and safe to eat. In Spain, mussels destined for foreign countries must by law be disinfected by holding them in tanks with a very weak solution of bleaching powder (sodium hypochlorite) for 48 hours. Another thing that puts people off is the frequent occurrence of tiny, gritty "pearls" in mussel flesh. These are caused by a parasite that enters the mussel while it is growing on the bottom. With off-bottom culture, the parasites cannot reach the mussels, and pearls are no problem.

Several growers are raising mussels on the coast of Maine, using raft culture. However, none of them are yet breaking even, and they must support their mussel growing by fishing and lobstering. As one of them ruefully told me, their best hope is for a large corporation to enter the business and buy them out or use them as subcontractors.

Clams on the half-shell, steamed clams, clam chowder, fried clams, and clam stew have long been American favorites. Abundant natural supplies have until recently made culturing clams an academic question. But scientists at VIMS and other institutions are now developing hatchery techniques to augment the seed stock of

A mussel-culture raft floats in the protected waters of Damariscotta Cove, Maine. Author could not learn the purpose of the conical object on top of the raft.

COURTESY OF UNIVERSITY OF NEW HAMPSHIRE—
MARINE PROGRAM, DURHAM, NEW HAMPSHIRE

clams, in addition to genetic programs to improve the breed and methods for protecting young clams against the horde of predators they face.

Clams are more mobile than mussels and oysters. Most species of clams dwell in soft sandy or muddy bottoms, in which they burrow with the aid of their hatchet-shaped feet. To move itself, the clam pushes its foot into the bottom and pumps blood into it. This makes the tip of the foot mushroom out and form an anchor. The clam then contracts the muscular stem of its foot and pulls itself forward. Clams can dig their way down at a prodigious speed when alarmed, as many an amateur clam-digger can testify. Later, they make their way up close to the surface again.

This buried existence shields clams from some of their numerous predators. The clams feed and breathe via a tough, muscular siphon,

162

A Maine mussel culturist shucks a mussel to inspect the quality of its meat. Raft-grown mussels are free from sand and "pearls."

COURTESY OF UNIVERSITY OF NEW HAMPSHIRE—
MARINE PROGRAM, DURHAM, NEW HAMPSHIRE

Thinning the luxuriant crop of mussels on a rope must be done periodically.

COURTESY OF UNIVERSITY OF NEW HAMPSHIRE—
MARINE PROGRAM, DURHAM, NEW HAMPSHIRE

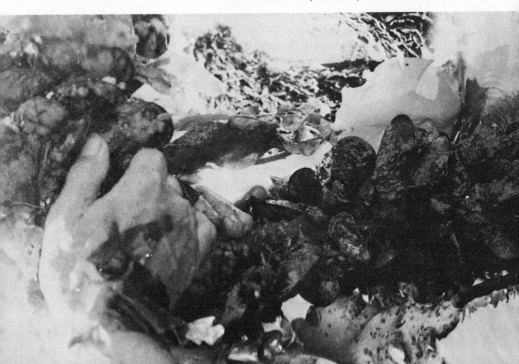

or "neck," that can be protruded above the bottom or retracted swiftly in case of danger.

There are many species of clams. The most commercially important species in the United States are the soft clam (*Mya arenaria*) and the hard clam, or quahog (*Mercenaria mercenaria*). Soft clams are eaten mainly steamed and fried; quahogs are consumed mainly raw on the half-shell and in chowder. Indians who lived on the East Coast used quahog shells to make the famous wampum—polished beads of shell that served as both jewelry and money. Hence the scientific name, which is ultimately derived from a Latin root meaning "trade" or "wages."

At the VIMS experimental station at Wachapreague, Virginia, hard clams are spawned in the same fashion as oysters, and the larvae are raised in tanks on a diet of cultured algae. The larvae set when they are eight or nine days old and the size of small grains of sand. They can be distinguished from sand by the fact that they do not swirl around when the water is stirred: they attach themselves to the sides and bottom of the tanks by byssal secretions. In a few more days the metamorphosis is complete, and the baby clams are scraped off the tank and transferred to grow-out trays. They are still so small that 250,000 clams fit comfortably into a two-by-eight-foot tray without touching each other.

In six to eight weeks the clams, which are the progeny of selected, fast-growing parents, reach two to five millimenters (about 1/12 to 1/5 inch) in diameter and are ready to plant out.

At this stage the juvenile clams are still highly vulnerable to predators, and the culturist's first task is to provide them with protection. One VIMS-developed technique is to plant the little seed clams in specially prepared beds of crushed stone. Beneath the stone fragments the clams are safe from such predators as crabs and small fish.

Another VIMS innovation, described by veteran scientist Mike Castagna as the real breakthrough, is a circular plastic-mesh baffle that goes around each planting bed and keeps the currents from sweeping away the baby clams before they have time to grow and burrow into the bottom. The baffles are set out in clusters of six, called rosettes, and the clams are planted inside them at an average

Top: At a National Marine Fisheries laboratory, a scientist compares gowth rates of hardshell clams and oysters grown in water of different temperatures.

Bottom: In a laboratory demonstration, a male hardshell clam, or quahog, emits a milky cloud of sperm. For actual spawning, groups of male and female clams are placed together in a water-filled tray.

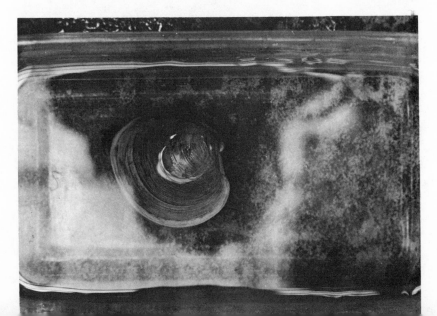

density of 250 per square foot. Each baffled area holds about three thousand tiny clams.

When clams reach an inch in diameter, they pass a turning point in their existence, for they are now too large for most crabs and fish to eat. However, they still have a major predator, the cownose ray, which roots them out of the bottom and crunches their shells between its flat, powerful teeth. The VIMS group developed an anti-ray fence made of the same plastic netting that fruit growers use to protect cherries from hungry birds. Sold in rolls eleven feet wide, the mesh effectively screens out the three-foot rays, which can wipe out a planting of clams overnight.

Working together with local clammers, the VIMS researchers discovered that a tent-like covering of mesh worked even better than an upright fence. The latest design uses floats to buoy the net up in the middle so that it won't settle on the clams and interfere with their vital water supply. Lengths of chain weight the ends down and anchor the whole thing in place. At harvest time, you simply flip the net back and dig away.

The fences and tents give a survival rate of about 70 percent or better; in one case as high as 86 percent. The clams grow rapidly, reaching the "New York nick" size* in as little as sixteen months, although the slower growers among them take twenty-four months or more. In the wild, it takes them three years or longer to reach this size. Thanks to their protection, the clams grow so thickly that they can be harvested with a potato rake. This is done twice a year at a spring low tide.

The only care the clams need during this time is maintenance of their protective netting and putting in a few crab traps. Crabs get in through the half-inch mesh of the netting while they are larvae, and when they grow they cannot get out again.

VIMS holds clam-growing seminars open to interested people from all states, and the techniques developed there have proven

---

*"New York nick" is a dealer's term for clams that measure at least one inch thick (through both shells).

One of the mesh baffles developed by VIMS researchers to protect baby clams from currents.

VIMS/SEA GRANT

useful in many areas. On Cape Cod, one town found that a four-by-eight-foot mesh pen planted with clams in a salt marsh produced sixteen bushels of littlenecks* in twenty-eight months. The cost of the pen, including materials, labor, and seed clams, was $278 (these were 1976 prices), and the value of the harvest was estimated at $1,000, a nice return on a virtually labor-free crop, even allowing for inflation.

A small group of culturists on Martha's Vineyard, backed by a consortium of five towns, are raising seed clams to plant in local beds with a variant of raft culture. They use floating sandboxes holding three to four inches of sand for the clams to burrow into. The boxes are suspended about eighteen inches below the surface of the water, and it takes an average of one growing season to get the clams to one inch, at which point they are planted out. (The growing season is from spring to fall. In theory the clams could stay on their sandbox-rafts forever, but year-round raft culture is not practical at the Vineyard because of the damage that ice does to the rafts.) Results have been encouraging, and the group plans to expand its operations. In 1979 they produced 1 million seed clams, and now they are experimenting with scallops as well.

Relatively little has been done in the culture of soft clams, other than thinning out overcrowded beds and spreading the surplus clams in new areas. However, certain coastal communities in Maine are now systematically rotating their clam beds, a first step in the direction of culture. Soft clams have been spawned artificially, but apparently it is not yet profitable to raise them for seed.

Clam culture on the West Coast and in Europe is still in its infancy, although the French government has discovered that French citizens will eat clams and is experimentally raising several European and American species. In Japan, however, clams have been raised for over one thousand years. Nine species of clams are grown commercially in Japan. The principal cultured species, which the Japanese call *asari*, is planted on beds that have been prepared by harrowing the seabed. This is done with a tractor at low tide, and it accomplishes the double job of loosening the bottom for the clams

---

*Littleneck clams are small quahogs, about one-and-a-half inches in diameter, except in Massachusetts, where they must be over two inches in diameter. Cherrystones are quahogs about three inches in diameter.

and getting rid of seaweed. *Asari* reach market size in one to two years. Starfish, octopuses, drills, and crabs eat a good many of the crop, but the worst predators are ducks.

In Taiwan, a species of clam known as the blood clam is raised for local consumption. Little care is given to the clams, although many growers fence them in to exclude predators and to keep the clams from escaping: the blood clam spins a long, sticky thread that water currents seize, thus wafting the clam away.

Elsewhere in Asia, cockles are very popular. They are cultured on a large scale in Thailand, Malaysia, Vietnam, and the Philippines. Related to clams, cockles look like clams with deeply cupped, heavily ribbed shells. Seed cockles are collected from the wild and sold to growers. Cockles need little care as long as they are in a good location. The grower's principal worry is poachers.

Scallops are one of the most popular seafoods in this country, so popular that they are often counterfeited. Many a "scallop" has actually been punched out of a shark filet with a cookie cutter. Genuine scallops, however, are bivalves, and they are the flighty gadabouts of the family. While oysters can't move at all, and mussels and clams can only crawl and burrow, scallops can leap through the water by clapping their shells together. This creates a jet of water with considerable propulsive power. Scallops also have a row of tiny blue eyes along the edge of their mantle. The eyes can detect light and dark, thus warning the scallop of the approach of an enemy (at least of one that casts a shadow). The scallop, alarmed, claps its shells and skips off across the bottom, a maneuver that often saves its life.

This behavior of clapping its shells has given the scallop a very large and powerful adductor muscle, which is the muscle that closes the shell. It is this muscle that we eat, discarding the rest of the animal. With clams, oysters, and mussels, on the other hand, we eat the viscera and discard the muscles.

Scallops are worldwide in distribution, and there are several hundred species. Several are fished commercially. In the United States the important species are the bay scallop, which reaches four inches in diameter and lives in relatively shallow water, and the much larger sea scallop, or deep-sea scallop, which reaches eight inches in diameter and lives at depths down to 300 feet.

Scallops can be spawned and reared in captivity, although so far this has only been done experimentally. Scallops are not yet cultured commercially in the United States or Europe, but in Japan they are raised in net bags suspended deep in the water. In southern California, researchers from the National Marine Fisheries Service and San Diego State University are experimentally raising a species new to commerce: the purple-hinged rock scallop.

This scallop, unlike most of its kin, lives free for only about six months. Then it cements itself fast to a hard substrate and stays there. The research group spawns adult purple-hinged rock scallops in the lab and raises the larvae in tanks. When large enough, the scallop larvae are transferred to floating pens made of plastic window screening and placed in the ocean, where they feed on natural plankton. The experimenters originally hung the scallop pens beneath an old cabin cruiser with a nonfunctioning engine but had to change their method when the ancient boat sank.

When they are about an inch in diameter, the juvenile scallops are glued to small plastic cages that are fastened to concrete or asbestos-board supports. In about a month, the scallops cement themselves to the hard support, and the cage is removed. The researchers estimate that their cultured scallops will take three years to reach market size, which is six inches across, with a muscle weighing one-eighth of a pound.

Man eats many kinds of water-dwelling snails, from the tiny winkle of Britain to the foot-long queen conch of the Caribbean. There is only one, however, for which the demand is so great and the supply so small that people are seriously interested in culturing it. This is the abalone.

Abalones do not bear much resemblance to other snails. They have flattened, roughly oval shells with a row of breathing holes along one side. The inside of the shell has a beautiful, iridescent luster, so that abalone shells have long been favorites of shell collectors and souvenir shoppers. It is the ab's meat, however, that makes it the object of man's desire.

There are about one hundred species of abalone distributed throughout the world's oceans, but only ten or so grow large enough to be commercially important. These species all live in temperate waters, and the major producing areas are Japan, China, South Africa, New Zealand, the south coast of Australia, and the Pacific coast of the United States, with southern California as the focus of the industry.

Eight species inhabit the Pacific coast of North America, of which three are fished commercially: the so-called red, pink, and green abalones. Actually most abalones' shells on the outside look a dull reddish-brown where they are not covered by algae, barnacles, and other fouling organisms. Furthermore, most species have several color variations. The easiest way to identify the species of an abalone is by the corrugations and whorls of the shell, the number and location of the breathing holes, and by the color of the animal's foot and the numerous mini-tentacles that fringe it. Red abalones reach a maximum size of twelve inches, pinks and greens about nine inches.

Young cultured red abalones, clinging to oyster shells, are about to be released into the ocean in a California state restocking program. Some bear stainless-stell identification tags.

COURTESY OF THE STATE OF CALIFORNIA DEPARTMENT OF FISH AND GAME

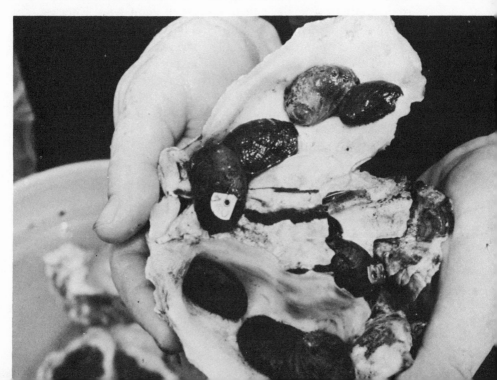

Abalones live from the intertidal zone down to depths of two hundred feet or even deeper. Due to overfishing, most of the remaining reds, pinks, and greens are found in the harder-to-reach depths. Abalones cling to rocks with their large, muscular foot and glide slowly along, grazing on algae, kelp, and other seaweeds. Mating takes place in shallow water. The male emits his sperm, and this chemically signals the female to release as many as 10 million eggs.

The eggs hatch in thirty-six hours, and the larvae swim free for six days, living on their stores of yolk. They then sink to the bottom and begin grazing on diatoms and other microorganisms. As is the rule in the sea, only a tiny fraction of all the eggs make it through the larval stage.

Abalones are harvested by divers who use a foot-long, chisel-pointed steel bar to pry the animals loose from their footholds on the rocks. In California, the pry bar has projections and notches that indicate the legal minimum sizes for red, pink, and green abs. Other than man, the only major predator of full-grown abalones is the sea otter, which knocks the abalones loose with rocks, a remarkable example of an animal's use of a tool.

Most abalone divers view the sea otters as rivals and hate them. But the sea otter, itself nearly exterminated by man's unbridled greed for its fur, is now protected by law. Sea otters also play a vital role in preserving the valuable beds of giant kelp. They eat sea urchins, which destroy the kelp's holdfasts. The kelp forms a habitat for many commercially valuable marine life forms and is itself a source of colloids that are used in a host of industries.

Between the divers and the otters, the native abalone population has dropped to the point where over half the annual United States consumption must be imported. The obvious answer is to culture the animals, but it has been a long time coming.

Abalones are difficult to raise through the egg and larval stages. In addition, they are finicky eaters. They will not eat just any old algae or seaweeds. They are quite definite in their preferences, and the preferred seaweed may not be easy to obtain. They also grow slowly: In the wild, it takes a red abalone about a year to reach one inch in diameter and four to eight years to reach legal minimum size (7¾ inches).

Japan is another nation where overfishing has severely depleted the abalone population. As early as 1959 Japan began a hatchery program to restock the depleted beds. The hatchery sells juvenile abalone from 0.6 to 0.8 inches to fishermen's cooperatives, and the fishermen plant them in protected locations such as rock crevices. Only about one percent of the larvae at the hatchery survive to the planting-out stage, and the fishermen recapture only 20 percent of that; however, the fishermen are reportedly quite pleased with the results. The restocking program now accounts for almost 5 percent of Japan's annual abalone catch.

Similar programs are under way in southern California, sponsored by Sea Grant and the California Department of Fish and Game, although as of this writing it is too soon to evaluate the results. Meanwhile, a culturist in Monterey, California, has been able to raise red abalone to market size in two-and-a-half years in an intensive system of his own devising. Patterned on the techniques used by the broiler chicken industry, the system is the result of seven years of intensive research into every aspect of abalone growth and reproduction. At each stage of development, the abalones live under optimum conditions of temperature, water quality, and diet. The culturist believes that by the mid-1980's he will be able to grow abs to market size in one year. He is already supplying thousands of baby abalones to a California state restocking program. Since California's strict regulations on abalone harvesting do not apply to cultured abs, a market may develop for small abalone that can be produced in a shorter time.

Rounding off the mollusk parade are the cephalopods, squid and octopus. These shell-less, tentacled creatures are delicacies in many Oriental and Mediterranean countries. Although they eat high on the food chain, they are now being considered as candidates for mariculture. Squid are not generally favored because of their habit of jumping out of their tanks. Octopus present certain problems, such as dependence on live food for their first few months, cannibalism, and a Houdini-like talent for escaping from confinement through unbelievably small apertures. However, octopuses have been reared experimentally to three-and-a-half ounces in six months and a

173

pound in nine months, and they command a fancy price in many countries; so perhaps Italian and Japanese gourmets may one day dine on cultured octopus. And the squid may be cultured in the United States and elsewhere as a laboratory animal, for its nervous system is important in the study of neurobiology.

CHAPTER **11**

Farming the Seaweeds

The casual visitor to the shore seldom realizes that seaweeds are anything more than Nature's beach litter. However, since ancient times man has used certain seaweeds as food, livestock feed, garden fertilizer, and medicine. In modern times they have beome sources of valuable industrial chemicals. In particular, seaweed colloids are used in hundreds of products, from ice cream to laxatives, from cosmetics to shoe polish, from vitamin pills to the specially formulated mud used in drilling oil wells. The colloids act as emulsifiers, thickeners, and binders.

Seaweeds are classed as algae, although they are much larger and more complex than the one-celled algae on which oysters live. There are three major groups of seaweeds: brown, red, and green. The browns and the reds are those of greatest use to man, although the green seaweeds *Ulva* (sea lettuce), *Monostroma*, and *Enteromorpha* are eaten in some parts of the world.

Seaweeds have a bizarre and complicated reproductive life, in which the familiar plantlike forms alternate with microscopic, one-celled, actively swimming forms, and sexual generations alternate with nonsexual. Not all the seaweeds follow this pattern, and the sex life of the red seaweeds is particularly complex. Only in the last

175

century have biologists really begun to solve the unknowns in seaweed reproduction; this knowledge has been of the greatest importance in promoting seaweed culture.

The major seaweed colloids are algin, carrageenan, and agar.
Algin is produced from brown seaweeds such as kelp and *Laminaria*
(often known by names like oarweed and sea girdle). It is used as an
emulsifier in ice cream and sherbet to keep the fats from separating
from the water; it also gives these products a smooth texture by
keeping the water from forming large ice crystals as it freezes. It is
used as an emulsifier in chocolate milk and in many other foodstuffs,
such as imitation whipped cream, jellies and jams, cheese, pie
fillings, baked goods, and salad dressings. It also gives beer a long-
lasting head. In the textile industry, algin goes into sizing and certain
dyes. It is used in printing inks and as paper sizing, and even in some
formulas for fireproofing textiles. And algin, blended with selected
clays and other minerals, keeps oil-drillers' mud flowing smoothly to
lubricate the drill head, seal cracks in the rock, and flush the cuttings
out of the hole.

In medicine, algin derivatives go into surgical gauze to make blood
clot quickly, into pills to let them break up quickly when swallowed,
and into various salves and creams. Dentists use an algin product to
make impressions of teeth.

Carrageenan was originally produced from the red seaweed *Chondrus crispus*, also called carrageen and Irish moss. Today it is extracted from a number of other red seaweeds. Carrageenan has been
used for centuries in western Europe and the United States to make
milk custards. It has also long been used in hand lotions and other
skin-care preparations. Today it is used mainly in the food industry,
particularly in dairy products and imitation coffee creamer. In bread
and other baked goods, it keeps the product moist. Carrageenan
goes into many dietetic foods. It gives body to frozen fruit juices,
pizza sauce, some baby foods, and creamed soups. Makers of frozen
vegetables use it in the little sauce packs that come in the package to
add flavor to the vegetables. A newer use is in dry, packaged mixes
for jello-type desserts and aspics. These mixes are useful in unin-
dustrialized countries because they keep for a long time without
refrigeration.

Over 50 percent of the world's toothpastes contain carrageenan as a binder; this useful colloid is also used in many medical and pharmaceutical products. In cosmetics, it goes into lotions and creams and shampoos. It keeps chewable pills from tasting too chalky; it keeps milk of magnesia from separating; and it is important in ulcer medicine and many liquid antibiotics. It is also finding use in medicine as an anticoagulant for blood.

Agar comes from several species of red seaweeds. The word comes from *agar-agar*, the Malay name for one of the agar-yielding seaweeds, although agar was apparently first used in China and Japan. One of agar's qualities is that it is practically indigestible by man. For this reason it has been widely used as a bulk-type laxative. Most of the bacteria that cause food to putrefy cannot digest it either; so agar is often used in tropical regions as a short-term food preservative. The same quality makes it useful in packing canned meats and fish. In the baking industry, agar is used in icings and glazes. It is also used in manufacturing photographic film, shoe polish, shaving cream, and hand lotions. A specialized use is as a lubricant for the dies through which tungsten wire is drawn for electric light bulb filmanets (for this it is mixed with graphite). But agar's most valuable use has unquestionably been as a culture medium for bacteria. Discovered in 1871 by the great German bactgeriologist Robert Koch, this application of agar has contributed immeasurably to medicine and science.

Seaweed culture is essentially a matter of providing a suitable substrate for the seaweed spores to settle on and develop into plants. However, seaweeds do not grow everywhere. They must have light in order to photosynthesize, and this limits them to the upper few hundred feet of the water. The further from the equator, the less penetrating power the sunlight has, and the depth limits of seaweed become shallower. Turbid water also screens out light very effectively. Seaweeds also require nutrients in the water, which rules out large areas where the water is nutrient-poor. Lastly, most seaweeds, particularly those in which man is interested, need a firm surface to which their rootlike holdfasts can cling. They are not adapted to a floating life, and they die when uprooted by storms. The notable exception is the famed sargassum weed of the Sargasso Sea. A similar species floats in the Gulf of Thailand.

Thus seaweed culture for all practical purposes is limited to the relatively shallow coastal waters of continents and islands, and even there local conditions of water temperature and quality dictate what species you can raise. Sewage has wiped out a number of once-productive seaweed beds.

It began in the Far East—China, Japan, and Korea—where since pre-Christian times certain species of seaweeds have been used for food and medicine. Seaweed culture (as opposed to simple gathering) goes back to the 1600's in Japan and perhaps earlier in China. Japan leads the world in seaweed culture today, and most of its production is still used for food.

The chief cultured seaweed in Japan is the broad-bladed red seaweed *Porphyra*, called *nori* in Japan and *laver* in the British Isles. Nori is the most valuable single sea product of Japan, and sixty to seventy thousand people are in the business of culturing it. More than one species of *Porphyra* is raised, and sometimes it is adulterated with the green, lettucelike *Monostroma*, which grows on the nori nets as an uninvited guest.

Nori likes cold water, but requires warm water for part of its life cycle. The traditional culture method, still used in Korea, was to set bundles of sticks in the water in early fall (September to October). Within two to four weeks, drifting spores of the nori weeds settled on the sticks and began to sprout into thalli (leafy blades). The sticks were then moved to another location close to shore, such as the mouth of a river, where the growing plants would get the benefit of dissolved nutrients brought down by the river. The blades were (and still are) harvested throughout the winter. By April the water starts to warm up again, and the nori blades begin to die off.

The crude bundles of sticks were long ago replaced by a variety of nets on which nori spores settle and grow. But until the mid-twentieth century no one knew where the spores came from or what happened to nori during the summer. Then in 1949 a British biologist, Kathleen Drew, discovered the missing piece of the puzzle: The plant spends its alternate generation as a small, filamentous alga that bores holes in mollusk shells. So different is this generation from its parent that seaweed specialists originally considered it to be a separate genus and gave it the name of *Conchocelis*. *Conchocelis* grows during the warm summer months. In fall, when the sunlight wanes and the water becomes chilly, it releases its spores, which in turn develop into nori.

Dr. Drew's discovery led to a revolution in nori culture; indeed, it proved so valuable that the Japanese government erected a statue in her honor. It enabled nori growers to develop an efficient technique of spore collection for the valuable seaweed. Most spore collecting is done at government laboratories. In early spring, strings of oyster shells are hung in large tanks of seawater and chopped nori blades are added. These release spores that settle on the shells and give rise to the *Conchocelis* generation. The shells are cultured in the tanks all summer. In autumn, the water is cooled to make the plants liberate the spores that produce the nori generation. For a fee, nori farmers soak their nets in the tanks to collect the spores and then string them out in the sea. The Japanese nori industry uses about three hundred miles of nets in all, and in places the poles that hold the nets bristle above the water like a forest.

The nori is harvested when its blades reach six to eight inches in length. If they are allowed to grow larger they are usually torn lose by waves. New blades grow after the harvest, so that one net can be harvested three or four times. Then the seaweeds lose their vitality, and the net must be replaced. Often seeded nets are frozen when their seaweeds are in the bud stage, and kept in storage so that they can be placed out in winter.

Nori is processed for the market by washing the blades in fresh water, chopping them into small pieces, and forming them into sheets that are dried. Other seaweeds eaten in Japan include *Mono-stroma*, which is seldom cultured, and the brown seaweeds *Laminaria* and *Undaria*, called *konbu* and *wakame*.

Large quantities of wakame are cultured in much the same way as nori, including artificial seeding of the spores. In parts of northern Japan, where there are large natural stands of wakame, intensive culture is not used, but the seaweed farmers toss clean, bare rocks and chunks of concrete into the sea to provide attachments for more spores. Konbu is not cultured on the same scale as nori or wakame, but its culture is increasing. Konbu growers often prepare the bottom by blasting it with dynamite, which also gets rid of competing seaweeds for a while.

Another Japanese seaweed that is cultured on an increasing scale is the red seaweed *funori (Gloeiopeltis)*. Funori is used for food and also for glue, sizing for silk, and hair-waving preparations.

China has specialized in cultivating *Laminaria japonica*, one of the Japanese konbu species. Used mainly is a floating-culture method in which the plants are hung from long ropes supported by bamboo floats. This method produces over 100,000 metric tons annually. In sterile waters such as the Yellow Sea, the growers hang porous clay jars filled with fertilizer from the ropes. Chinese seaweed scientists are conducting a genetic program to create new strains and hybrids with greater productivity, higher iodine content, and lower water content. To hasten the process, they create mutants by irradiating spores with X rays.

Chondrus crispus, or Irish moss, the main source of carrageenan, grows naturally off the coasts of the British Isles, New England, Canada's Maritime Provinces, and Korea. It is not at present cultured. "Mossers" gather it by hand, either by raking it loose from underwater beds or gathering the weeds from the beach when they are washed up by summer storms. However, Marine Colloids, one of the world's largest seaweed-colloid producers, is culturing Irish moss experimentally at a pilot plant in Nova Scotia and has plans for an installation to raise one million pounds a year in plastic-lined ponds.

The same company already has a sizable seaweed-culture program in the Philippines, where family farmers raise the red seaweed *Eucheuma*. This warm-water seaweed is also rich in carrageenan. Small sprigs of *Eucheuma* are tied to nets with a one-foot mesh, supported by mangrove stakes. The plants sprout and grow vigorously, and they are ready to harvest in about sixty days. Thus a seaweed farmer in the Philippines can harvest six crops a year. The nets sit about three feet off the bottom, just below the low-tide level, to keep the plants covered with water at all times.

The plants are harvested by swimmers. About half a pound of plant mass is left at each planting point so that it will produce new blades and replanting is not necessary. An acre of *Eucheuma* yields about 40 metric tons of fresh seaweed a year, equivalent to 5.3 tons of the dried weed. *Eucheuma* is also eaten as a vegetable in the

Philippines, lightly boiled and seasoned with hot pepper, soy sauce, and lime.

One seaweed crop to which much attention is now being given is the giant kelp *Macrocystis pyrifera,** native to the Gulf of Alaska. *Macrocystis* is a cold-loving member of the brown seaweed family, and it commonly grows in water from twenty to sixty-five feet deep. Where the water is extremely clear, it is found as deep as one hundred and twenty feet. Anchored by masses of tough, rootlike holdfasts, it sends its mighty stems up toward the surface, growing at a rate of up to two feet a day. Long fronds sprout from the apex of the holdfast and are buoyed by gas-filled float bladders. When the fronds reach the surface, they continue to grow, sprawling out along the water and producing a thick canopy. This is a very efficient arrangment for trapping sunlight and probably accounts in part for the rapid growth and vast size of giant kelp. *Macrocystis* plants reach one hundred feet in length, and individual plants have reached two hundred feet. Plants may survive many years, though individual fronds live only about six months.

A kelp forest is a unique ecosystem. The kelp provides shelter and food for hundreds of organisms, including sea otters, abalones, and commercially valuable species of fish. Fish and invertebrates graze on the kelp fronds, and many small organisms attach themselves to the kelp for life. The only real enemy of giant kelp is the voracious sea urchin, which nibbles off the holdfasts.

The kelp industry of California dates back to World War I, when imports of potash from Germany were cut off. Since our farms required thousands of tons of potash for fertilizer, some enterprising soul turned to harvesting kelp for its potash content. The kelp was burned and the potash extracted from the ashes. The important industrial solvent acetone was also obtained from kelp, in this case by fermenting it. With mechanical harvesters, kelp production rose to nearly 400,000 tons a year. During the war and the years that followed, many uses had been discovered for algin, in which the kelps are rich, and the great harvests continued.

An ingenious mechanical harvester was developed to gather kelp.

---

*Another giant kelp, *Nereocystis*, grows from California to Alaska. It reaches a maximum length of about one hundred and twenty feet. Elk kelp, or *Pelagophycus*, occurs from southern California as far south as central Baja California and reaches lengths of one hundred and twenty feet.

Towed by a powered barge, it glided through the water three or four feet below the surface, snipping off the kelp with a reciprocating cutter bar like a farm haycutter. Another mechanism grabbed the cut kelp and loaded it on board. In improved form, the same device is used today.

With the mechanical harvester, kelp production rose to nearly 400,000 tons per year of the wet weed, or an average of three and a half 300-ton barge loads a day. After the war, the market for kelp potash declined, but meanwhile uses had been discovered for algin, and the great kelp harvests continued.

If cropped near the surface, giant kelp regenerates rapidly and can be harvested two or three times a year. However, a variety of factors, including sewage pollution from the burgeoning coastal communities, heavy grazing by sea urchins, and a two-year "heat wave" in the water, killed off 90 percent of the great California kelp beds. By 1960, the giant kelp was at its lowest point in recorded history.

To the rescue came scientists from the California Institute of Technology and the Scripps Institution of Oceanography, working with Kelco, a large producer of algin. Research on methods to restore the kelp beds was soon underway.

The first technique they put to use was controlling sea urchins. This is done by divers who either smash the urchins with hammers (effective but slow and costly) or douse the sea bottom with quicklime, which poisons the urchins just as it does starfish. The researchers learned that wiping out the urchins did not always guarantee a regrowth of giant kelp. Other seaweeds often took over the sea bottom and shaded out the *Macrocystis* seedlings. Apparently *Macrocystis* spores can only germinate successfully within a few hundred feet of their parent plants. By limiting the urchin clearance program to a small radius around surviving stands of adult *Macrocystis*, the researchers got stands of young kelp. When these were established, the divers would then clear the sea urchins from another strip of territory, gradually expanding the kelp stands.

Then came the problem of what to do about areas where *Macrocystis* had vanished and there were no spores to produce new plants. The solution was to thread a stout nylon cord through the holdfasts of growing kelp plants, then pry the holdfasts loose from their attachments. The diver who did this job then fastened the cord to a tow rope. When the rope was full, a boat towed the plants to the new location. Since the maximum safe speed for the kelp is one mile an hour, for long trips the plants were loaded onto the boat and kept moist by periodic dousings with seawater. At the transplantation site, the kelp plants were fastened to small buoys attached to a heavy chain, to rocks, or to other supports. The kelp's holdfasts soon provided a firm anchorage. Both young and mature plants were successfully moved by this method.

Another technique involves culturing *Macrocystis* embryos in the lab and spreading them over a suitable bottom area. Since only about one embryonic sporophyte of every 100,000 succeeds in attaching itself and developing into a juvenile kelp plant, it is necessary to seed the area with astronomical numbers of kelp embryos. Dr. Wheeler North, one of the chief scientists of the project, reported that he and his team had to develop a method for producing several billion kelp embryos at a clip, every ten to twenty days. The microscopic embryos are poured down a hose that extends to the seafloor to keep them from being carried off by currents or devoured by plankton eaters on their way to the bottom.

Research has also been completed on culturing warmth-tolerant strains of *Macrocystis* from Mexican waters to be grown near the discharge pipes of power plants. Since the coastal areas where giant kelp thrives are limited, a bold scheme for growing kelp in the open ocean is planned. It involves anchoring kelp plants to lines on frames that look like giant whirligigs of the type used for drying laundry. The frames, supported by buoys, are anchored in deep water. They lie from forty to eighty feet below the surface of the water, a depth readily reached by sunlight in the clear waters of the open sea. Since the open sea is low in nutrients, a wave-powered pump creates an artificial upwelling of cold, nutrient-rich water from depths of one thousand feet or more. A pilot project off San Clemente Island gave encouraging results.

Spaced ten feet apart on the lines, 436 kelp plants could be raised on one acre. Plans for the first full-scale open-sea kelp farm envision 100,000 acres. The vast quantities of kelp that will be produced can be converted into methane, for use as a fuel to replace dwindling oil resources; animal feed; liquid fuels; colloids; and other industrial products. As a fringe benefit, it is expected that edible fish will thrive among the kelp. If the project sounds visionary, remember that it is proposed by a respected Navy scientist.

CHAPTER **12**

Some Final Thoughts

What will the future bring for aquaculture? Predictions are risky when so many variable factors are concerned, but the indications are that aquaculture will expand as the natural stocks of fish and shellfish become scarcer and costlier to harvest.

Intensive culture of trout, catfish, and other "luxury" species will probably become more intensive as improved technology of supplying oxygen makes it possible to raise more pounds of fish per gallon of water. One constraint on expansion is a growing shortage of water, and the EPA predicts that the situation will become much worse. This will probably spur efforts to develop commercially practical recycling systems, in which the water is used three or four times before leaving the system. This will make necessary much more efficient and cheaper filtering systems.* However, even the best recycling systems need constant replenishment because they are constantly losing water by evaporation. Mechanization and automation will increase because they are cheaper than hand labor.

In the meantime, countries such as Chile, Argentina, and New Zealand, which have abundant supplies of pure, cold water in their lofty mountains, may go into large-scale culture of trout as an export item. As a related specialty, they may produce eggs so that trout

184

farmers in the Northern Hemisphere can extend their operations into what is now the off season. (Spawning time in the Southern Hemisphere is halfway around the year from that of the Northern Hemisphere.)

Salmon ranching is another promising possibility for South America. Domsea Farms, a West Coast salmon grower, has gotten exciting results from a pilot salmon-ranching project in southern Chile. When released, the smolts swam poleward to the Antarctic Ocean and feasted on the rich supplies of krill. They returned fat and healthy, and matured in about half the time it would have taken them in the North Pacific, their native environment.

The patient work of fish geneticists in developing improved strains of trout, catfish, carp, and other cultured species will begin to show results. Right now, fish geneticists are about at the same point as chicken breeders were 150 years ago. The development of pure, high-yielding strains of fish analogous to the White Leghorn or Rhode Island Red chicken offers exciting possibilities. (The same holds true of mollusks and crustaceans.) It is not a speedy process, however. With most species, fish geneticists can count on a wait of about five to seven generations, covering ten to fourteen years, before results begin to show. For late-maturing species, such as sturgeon, the waiting period is correspondingly longer; for early-maturing species it is less. Cultured tilapia can be spawned at sixty to ninety days of age, and Mike Sipe in Florida has been able to produce seven generations in thirty six months. However, tilapia are exceptional in this respect among cultured food fish.

Aquaculture experts are showing new interest in polyculture. At present, polyculture is used mainly in China, where it is highly developed, and in low-intensity, subsistence-type aquaculture elsewhere. However, studies at several American universities have shown that polycultures outyield monocultures significantly, at little or no extra expense. One experiment at Louisiana State University involved raising channel catfish in floating cages as the main species.

---

*In theory, the ideal solution would be a system in which algae, vegetables, or other plants extracted nitrogen and phosphorus wastes from the water and yielded a useful by-product. However, such systems require a tremendous amount of space and so far have not proven very effective in practice.

Roaming the pond freely were small numbers of buffalo fish and paddlefish, plus a seed population of crayfish that multiplied rapidly. The catfish were fed a standard pellet ration. The other three species lived on the feed the catfish wasted, plus natural plankton and other organisms nourished by catfish "manure." At the end of the season the catfish yield was equivalent to 3,300 pounds per acre, while the buffalo fish yielded 300 pounds per acre and the paddlefish 120 pounds per acre.* The crayfish, stocked at a modest rate of 25 pounds per acre, yielded 1,100 pounds per acre over a six-month trapping season. For comparison, a commercial catfish grower is happy with 2,000 pounds an acre, and a crayfish farmer is happy with 1,000 pounds an acre. As a side benefit, the buffalo fish, paddlefish, and crayfish served as an unpaid cleanup squad.

Both buffalo fish and paddlefish were cultured in the earlier years of this century, but they could not compete on the market with the much larger and cheaper natural catch. The buffalo (three species) is a large member of the sucker family, reaching fifty pounds and more in nature. It is found from Alabama and Texas north into Canada. The bizarre-looking paddlefish is named for its long, paddle-shapped snout, with which it stirs up the mud to get at the tiny animals it feeds on. Native to the Mississippi River drainage system, the paddlefish occasionally reaches ninety pounds. Its flesh is good, and it also yields an excellent grade of caviar. These and other native species may have a resurgence as auxiliaries in polyculture.

Most new commercial-scale enterprises in aquaculture, in the United States at least, will be backed by large corporations. Few private individuals have the resources to start such an enterprise on their own or to survive the possible loss of a year's crop to disease or the wrong kind of weather. However, many established firms are now finding it convenient to expand their output by subcontracting it to small growers. Thus, the big trout farms in Idaho's Snake River Valley have subcontracted thousands of tons of production to local farmers and retired people who raise the trout as a sideline or a money-making hobby.

---

*The catfish were stocked at the rate of 2,700 fingerlings per acre, the buffalo at 100 per acre, and the paddlefish at 20 per acre.

To succeed in commercial aquaculture, one should have a good, basic understanding of biology or, better yet, a background in fisheries biology, including a knowledge of fish behavior. A lack of such knowledge can have disastrous consequences, as in the case of the engineer who designed a silo-type tower for raising trout. Water was pumped in at the bottom, the idea being that this would move the fish up to the top of the silo for easy harvesting. However, the natural response of trout when alarmed is to swim against the current. And that was what the trout did at harvest time, huddling stubbornly at the bottom of their tank. An electric grid at the bottom of the tank was turned on to drive the fish to the top. It failed ignominiously: the fish obeyed their built-in programming, swam against the incoming water, and were electrocuted. Needless to say, they should have been delivered to the processing plant alive, not as corpses.

For success in aquaculture, it is also important to have some knowledge of business procedures and record-keeping, or else to have a partner or backer who does. Most important of all is practical, on-the-job experience in raising the kind of organism you are interested in.

There are bright prospects for another type of aquaculture: subsistence aquaculture in tropical and semitropical regions of Latin America, Africa, and India. The stars of this endeavor will almost certainly be tilapia and carp because of their low cost to raise, fecundity, and hardiness. However, previously unutilized native species may well be brought into culture. Using land that is unsuited for conventional farming and relying on peasant communities to supply their own labor forces for the local ponds, this approach could do much to alleviate the protein shortage that is such a plague in poor countries. A number of governments are supporting pilot aquaculture programs. Several private philanthropic groups are also teaching peasants how to construct ponds and care for the fish.

Prospects appear bright also for small-scale aquaculture in the United States and other industrialized nations. The New Alchemy Institute, a group of privately funded researchers* based in East Falmouth, Massachusetts, have developed a number of different schemes for raising food fish. Believing that "small is better," they

Villagers in Nepal excavate a fish-culture pond under guidance of the FAO. Dirt removed from each section of the pond floor is stacked in uniform-sized mounds as a rough-and-ready leveling measure. The dirt will be carried away before the pond is flooded.

WFP/FAO PHOTO BY E. WOYNAROVICH

have designed their projects to use a minimum of energy and mechanical equipment. They make maximum use of solar energy and biological processes, as in their "solar algae pond." This is actually a cylindrical, translucent fiberglass tank in which algae and tilapia are grown together. A reflecting wall on the north side of the tank increases the amount of available sunlight. The algae supply most of the tilapia's dietary needs as well as oxygen; the nitrogenous wastes of the tilapia nourish the algae. Additional protein comes from earthworms that thrive in the New Alchemists' compost piles and from midge larvae that are cultured in shallow ditches.

The New Alchemists have also integrated fish tanks with greenhouses, a concept that is finding increased use. The greenhouse effect warms the water in the fish tank; the fish tank in turn acts as a heat sink, releasing heat to the greenhouse at night. The fish wastes make an excellent fertilizer for the greenhouse vegetables; alternatively, they can be added to compost piles. The vegetables wastes can be fed directly to the fish—this system is designed for herbivorous or omnivorous species—or composted to raise earthworms.

As its advocates point out, the integrated fish-tank-plus-greenhouse plan can be used by homeowners to produce a substantial proportion of their own food, even well into the winter. It can even be used by slum residents if they can get the use of a vacant lot or yard and protect their installation against vandals. A serviceable greenhouse and fish tank can be built simply at very little cost, using some scrap materials.

Another pioneer in the field of small-scale farming is the Rodale organization, a highly successful business firm. Founded in the 1940's by the late J. I. Rodale, an organic-food enthusiast, the firm publishes popular magazines an organic gardening and farming and operates an experimental farm near its headquarters in Emmaus, Pennsylvania. Part of the experimental program involves aquaculture.

*Most of the New Alchemmists' operating budget comes from small contributions from private citizens who like the work they are doing. A small amount comes from foundations and government agencies. The New Alchemists' main objective is to develop environmentally sound methods to enable people to live comfortably but not destructively, and to increase their self-sufficiency. One of their major projects has been an aquaculture project in Costa Rica. They also have an ongoing program of organic farming, developing biological pest controls and organic fertilizers.

On the experimental farm the Rodale researchers raise tilapia, carp, and channel catfish in nothing more elaborate than ordinary plastic wading pools, readily available at stores. These fish are warm-water species, and their normal growing season for that area is approximately June through October. But by covering the pools with plastic "greenhouse" domes, the Rodale team can now stock their pools with fingerlings in mid-April. Algae supply much of the needed oxygen as well as food for the fish (the catfish, being carnivores, are fed a protein ration). In place of the bulky standard biological filter a new device called a rotating biological contactor is used.

This is essentially a thick sandwich of corrugated fiberglass plates with an axle through the center, mounted on a plastic float in the pool. A small electric motor rotates the plates at a leisurely six revolutions per minute. Bacteria growing in the corrugations of the plates convert nitrogenous fish wastes into harmless compounds, and the constant turning of the plates keeps the water circulating. It has been found that the device also provides ample oxygenation, and the purified water trickles back from this height as a small "waterfall." The drop gives it enough force to stir up the water and aerate it.

The Rodale researchers have raised fifty pounds or more of edible fish per pool, starting with five-inch fingerlings. By keeping a pair of chickens in a cage over one pool, an experiment inspired by the Oriental practice of fertilizing fish ponds with manure, they were able to boost yields higher still. The ponds looked dreadfully unattractive, but the fish were not only perfectly safe to eat but of excellent quality.

The growth of aquaculture will stimulate the growth of a host of auxiliary industries such as the fabrication of raceways, tanks, incubators, filters and pumps, aerators, and nets for harvesting. There will also be a need for a greatly increased supplies of food organisms eaten by larval fish and crustaceans, such as *Artemia* (brine shrimp), *Daphnia*, and rotifers. There may well be a need for specialized producers to supply algae cultures to oyster hatcheries. Even such a highly specialized product as hormones to stimulate spawning is in demand right now.

In the foreground is one of Rodale's rotating biological filters, a thick sandwich of corrugated fiberglass plates. in the background, serving another tank, is a fountain-type aerator.

COURTESY OF RODALE RESOURCES, INC.

One field of opportunity that has barely been touched is the medical care of fish. At present the number of fish pathologists in the country is totally inadequate for the needs of fish farmers. One answer to this is the diagnostic kit developed by Rangen, Inc., a leading manufacturer of trout feed. The lunchbox-sized kit contains equipment and instructions for taking smears, cultures, fixing tissue samples on slides, and so on. The farmer mails his samples and about six of the affected fish to Rangen's well-equipped pathology lab in Idaho and receives a diagnosis and recommendations for treatment. Since delay can be fatal, and the mails are so often slow, Rangen is considering plans to open regional labs in other parts of the country.

Another promising Rangen project is a computerized program for hatchery management. This can be set up so that the hatchery manager calls the machine on his office telephone to get instructions and advice. The computer asks him certain questions, for example: How much feed did he give the fish in each raceway yesterday? How many fish died? How many pounds of fish did he ship out from each pond, if any? The manager types in the answers, and the machine prints out a new feeding schedule. It can also spot signs of disease problems, and tell him when to take samples for diagnosis. The computer can be programmed to order feed automatically and even to order eyed eggs or fingerlings for the next crop of fish.

Aquaculture may one day find large-scale application as a profitable method of handling waste water and sewage. The heated effluent from power plants and factories is already being used in raising fish and shellfish, and this use will almost certainly expand. In Israel and Germany, fish have been raised in lagoons of treated sewage.

The boldest such scheme, however, is the brainchild of Dr. John Ryther of the Woods Hole Oceanographic Institution. Ryther's original idea was to develop a sewage-purification method for seaside towns that would yield salable by-products. This would not only prevent the growth of repulsive algae blooms in coastal waters that would drive tourists from the beaches; the profits from by-products would pay for much of the cost of treating the sewage.

The system is essentially simple; a pilot project at WHOE's Quissett Campus worked beautifully. Treated sewage effluent from a nearby community was trucked to the campus and dumped into shallow, plastic-lined ponds where it was mixed with seawater. Marine algae were cultured in these ponds to use up the nutrients in the sewage—the nitrates and phosphates that cause algae blooms and eutrophication.

This still left Ryther with a lot of algae to get rid of, so he added oysters and clams to eat the algae. These mollusks lived in racks in concrete raceways, and the algae-laden water was pumped slowly past them. Although much of the nitrogen and phosphorus in the sewage was by this means converted into oyster and clam meat, the mollusks themselves excreted a good deal, and Ryther added another set of raceways where he grew seaweeds to take up the last of the nutrients. The seaweeds could, of course, have been grown directly in the diluted sewage, but Ryther was trying to create a polyculture system.

In fact, he added marine worms to eat up the solid wastes given off by the mollusks, and fish to eat the worms. He even added a few lobsters to see what would happen, and found that they grew at a rate almost double that of wild lobsters.

When the experiment was terminated because the funding ran out, two problems remained to be solved. One was the presence in the sewage effluent of pathogens that might be taken up by the animals being cultured in the system. This could be dealt with by depurating the fish and shellfish at harvest time, a well-established procedure.

The other was the accumulation of heavy metals and organic trace compounds, occurring in the sewage and potentially toxic to human beings, in the fish and shellfish. This was later found not to be a serious problem if purely domestic sewage were used, as it contains relatively low levels of such contaminants, unlike industrial waste-water. So seaside and lakeshore communities may some day put his thrifty technique of sewage reclamation to use.

To paraphrase a statement made by the FAO Technical Conference on Aquaculture in 1976, aquaculture is an efficient means of converting low-grade food materials and waste products into high-grade protein for human use. It can often be combined with agriculture and livestock farming, thus contributing to integrated rural

193

development. In its hatchery aspects, it can augment depleted wild stocks of fish, mollusks, and crustaceans for both commercial and sport fishing. Imaginatively planned and intelligently applied, aquaculture provides a means of creating jobs and revitalizing rural life.

It is to be hoped that governments will heed these findings and give their full support to this valuable industry. Aquaculture will not cure all the problems created by man's greed and destructiveness, but it will go a long way toward helping them.

Suggestions for the Do-It-Yourself Fish Farmer

This appendix is intended for the reader who wants to try his or her luck at raising a small amount, fifty to one hundred pounds, of fish at home. This amount would supply an average family of four with at least one meal of fish a week, and it can be raised with a minimum of investment in equipment. Experts agree that even the most enthusiastic beginners should start small and expand their operations (if they want to) after they have gained practical experience. Various government agencies publish simple pamphlets on fish culture. Get some and study them.

Remember that most states' fish and game laws place restrictions on aquaculture; so be sure to check with your local fish and game department before you begin. In many states this function now falls under another title, such as Department of Environmental Protection or Department of Conservation. If in doubt, check with your local public librarian.

In general, you can do what you want as long as your aquaculture installation does not connect with any public waters (this ensures that undesirable fish won't escape into streams and lakes and ruin the sport fishing there). However, in some states you could technically violate the fish and game laws merely by giving a fish to your next-door neighbor. So, once again, protect yourself by checking with the authorities beforehand.

Ponds or Tanks?

Constructing a pond is expensive and in most states requires a permit from the state authorities. Your local cooperative extension agent can tell you to whom you should apply. You may need a permit from your local zoning board as well. For these reasons, as well as those mentioned above, it is suggested that you begin your career as a fish farmer with a modest-sized container, not a pond. You will want a recirculating water system to avoid confrontations with the law as well as to economize on water, an important consideration when many communities ban the washing of cars or watering of lawns during water shortages.

A very convenient and easy-to-obtain fish tank is an ordinary vinyl wading pool. A pool twelve feet in diameter and three feet deep will hold approximately 2,500 gallons,* which is ample for as much as one hundred pounds of fish as long as you filter it adequately. Without filtration, a pool of this size can support about twenty pounds of fish. If your vinyl pool is new, fill it with water and let it stand a few days. Empty it and refill with new water. This procedure leaches out toxic chemicals that may harm the fish. If you enjoy working with your hands, you can build a serviceable tank yourself from plywood, coated inside with epoxy resin to waterproof it. Don't use creosote or wood preservative. They contain toxic chemicals that will leach into the water and kill the fish. A fish pool can also be built of cement block coated with epoxy or waterproof paint.

Many small-scale fish farmers use an empty oil drum or other metal drum, painted inside with epoxy to protect the metal from rust and the fish from heavy-metal poisoning. (Iron in the water reacts with the mucous coating of the fishes' gills to form an insoluble precipitate that suffocates the fish.) An oil drum holds fifty-five gallons and can accommodate as much as fifteen pounds of fish. This translates into twenty or so fair-sized trout or catfish. (A twelve-inch trout or cat weights about one half pound.)

*One U.S. gallon equals 231 cubic inches.

Your fish tank can be of many materials and dimensions, but whatever you do, remember that your fish need a minimum depth of eighteen inches of water.

Building a Filter

In addition to your tank, you'll need a filter. The filter's capacity should be fifty-five gallons for every one hundred pounds of fish you plan to raise. A good filter medium is crushed limestone of two-and-one-half-inch size.

Mount the filter on supports so that its bottom is a foot or two higher than the top of the fish tank. This allows you to use gravity feed to return the filtered water to the tank: it also provides supplementary aeration as the water splashs down. Be sure the supports are strong, for a filter this size is heavy. Enough limestone to fill a fifty-five gallon container weighs about half a ton! Of course, you are not limited to a metal drum for your filter. A waterproofed wooden box, an old bathtub, even a series of laundry buckets will serve.

Getting Ready for the Fish

About four weeks before you plan to stock your fish, fill your tank with water and set up your pumping system and filter. To spread water evenly over the top of the filter, have the pump feed into a rosette sprinkler nozzle, a section of old hose with holes punched along its length, or whatever else your ingenuity may suggest. If you are using a fifty-five-gallon drum for your filter, a handy water spreader is a ten-inch pie tin with numerous ⅛-inch holes drilled in it.

Start the pump and keep the system running. The purpose of the two-week running-in period is to condition the filter—that is, to give the beneficial denitrifying bacteria time to become established. To encourage their growth, add five drops of household ammonia to the water each day. Do not be overenthusiastic about adding ammonia. It is one of the main waste products you must get rid of once your fish are in residence.

Run the pump just fast enough to provide a slow, steady trickle of water. When the tank has fish in it, the filter may also screen out solid wastes. If not, vacuum or siphon them out once a week.

Aeration
Aerators come in many sizes and designs. You will need one of them. A 1/20 horsepower aerator should provide sufficient dissolved oxygen for one hundred pounds of fish. Be sure your fish have enough oxygen at all times. Lack of oxygen is the leading cause of death for intensively cultured fish. If the fish don't suffocate directly, they are so weakened that they fall easy prey to diseases and parasites. If you notice the fish staying near the surface, it is a sign that oxygen is running low. If they break the surface to gulp for air, it is an emergency signal.

If your tank contains a healthy growth of algae, you will usually not need aeration during the sunlight hours. On very hot days you may need to aerate anyway—the warmer the water, the less dissolved oxygen it can hold. *Green* algae are desirable. *Blue*-green algae give fish an unpleasant flavor. If you keep clear water in the tank, you will need aeration around the clock.

What to Test For
Water quality is important. You will need a simple kit to measure the pH of the water. pH is a measure of acidity and alkalinity, and the pH scale runs from 1 to 14. The lower the pH number, the more acid the water is; the higher the pH number, the more alkaline. The neutral point is 7.0. Fish thrive best between pH 6.5 and 7.5, and most species prefer their water slightly on the alkaline side, about 7.1 or 7.2. (You need not be so exacting, however, as long as the fish are healthy and growing.) Fish also do better in water that contains enough calcium and magnesium ions to make it slightly hard. A sprinkling of ground limestone helps neutralize acidity and also supplies the necessary hardness.

In addition to a pH testing kit, you will need a kit to test for ammonia. Dissolved ammonia should not exceed one p.p.m. (parts per million). Beyond that it becomes toxic to fish.

An oxygen meter is expensive but good to have. It saves guesswork and it saves fish. Dissolved oxygen (DO) levels should not be permitted to drop below 5 p.p.m.

Stocking Fingerlings

Now, at last, we come to stocking the tank. The species of fish you choose should be one that thrives in your climate. For areas with cool summers, trout are a good summer crop. Otherwise catfish, tilapia, or carp (where permitted) are a better choice. In parts of the South, some amateur aquaculturists raise catfish in summer and trout in winter. The same can be done in the North if your fish tank is combined with a greenhouse.

Plan on stocking your fish when reliably warm weather arrives. If you are a gardener, this is the time when you set out tomato plants. The growing season lasts five to six months in most parts of the United States.

Since you'll want to harvest your fish at the end of one growing season, and avoid the expense and bother of carrying them through the winter, you'll do best to start with good-sized fingerlings, four to five inches for trout, three inches or more for tilapia. You'll pay more for these fingerlings than you would for smaller fingerlings or fry, but you'll make up for it with more pounds of fish at harvest time. In addition, larger fish tend to have a greater proportion of edible meat. Order your fingerlings from a reputable hatchery (see the partial listings on pages 205–209).

Your fish will arrive in water-filled plastic bags in a chilled container. Don't give way to excitement and dump them into your tank immediately. It is important to bring the fish gradually to the same temperature as the water in the tank. This avoids thermal shock, since fish are extremely sensitive to sharp changes in water temperature.

There are two ways of doing this. One is to float the bags, unopened, in the tank for an hour. A better way is to pour out about half the water in the bag and slowly add enough water from the tank to refill it. Repeat this process several times, until the water temperatures have been equalized. Check the temperature in the bag and in the tank with a thermometer, another indispensable piece of equipment. When the temperatures are equal, gently release the fish.

Feeding

Unless you are a confirmed experimenter (and willing to sacrifice fish in your quest for knowledge), stick to commercial fish chow and leave homemade rations alone. Exception: Tilapia do well, especially as they approach maturity, on leafy vegetable scraps and grass clippings. But make sure that these greens have not been treated recently with chemical pesticides. Earthworms can substitute for a portion of commercial fish chow, but are not proven as a complete ration.

The New Alchemists found a handy way of dispensing earthworms to their fish. They place the worms on a flat Styrofoam float with holes drilled in it. Instinct drives the worms down the holes, and the fish pick them off on the bottom side.

Feed your fish about 3 percent of their body weight a day, and increase the ration once a week. With healthy fish and a satisfactory water temperature, you can assume that the fish gain one pound for each two pounds of food they consume (a conversion ratio of 2 to 1). You can estimate the approximate weekly growth of the fish by multiplying the amount of feed you gave them that week by 0.5, then adding this figure to the estimated weight of the previous week. (This means you must keep a careful record!) Multiply the result by 0.03 to get the next week's food ration. For simplicity, round off the figures to the nearest 1/100 pound.

For example, if you stock ten pounds of fingerlings, give them 0.3 pounds of feed each day for the first week. The next week, increase the feedings to 0.34 pounds a day, the third week, 0.37 pounds a day, and so on.

Feed the fish at the same time every day, and check to make sure that no food is wasted. Uneaten food decays and makes the water bad. If you use a floating feed, it is easy to tell how much is being eaten. Remember that it is better for the fish to underfeed than overfeed. Remove uneaten food as completely as possible.

Daily Care

Check your fish twice a day. Remove those that die. Watch to make sure that the fish are eating normally and do not show signs of disease. Be sure that the water level does not drop more than an inch or two, and top up the tank as needed. If your water supply is chlorinated, let the water stand overnight in a bucket to let the chlorine pass off. Once a week drain of the bottom few inches of water, where the impurities tend to collect. Refill the tank each time.

If you stock your fingerlings very densely, you will have to thin them out regularly to keep their biological demands from outgrowing the capacity of the tank. You can eat them yourself, dig them into the garden, or feed them to the cat, but if you fail to thin the fish population, you'll end up with a tank full of runty, undersized fish. Try not to exceed the recommended density of two pounds of fish per cubic foot of water. (A cubic foot is just under seven and a half gallons—7.48 U.S. gallons, to be precise.)

Fish need daily care. You cannot neglect them and expect good results. In the wild they can fend for themselves, but in intensive culture they depend on you. If you have a trip planned, arrange for a friend or neighbor to care for the fish while you are away.

You can harvest your fish all at one time by simply draining the tank. However, cleaning fifty to one hundred fish at one time is a lot of work, and the fish take up a lot of space in the freezer. Consequently, many home aquaculturists find it easier to harvest the fish a few at a time when they want a fish dinner. A simple dip net is used for this purpose.

Variations
If you have no room in your yard for a fish tank, you can still raise fish in your cellar. Researchers at Rodale have raised catfish to one-pound size and even larger in basement pools measuring eight feet square and eighteen inches deep. Just make sure that your basement floor is strong enough to take the added weight: this much water weighs nearly six thousand pounds. (See page 202.)

Putting the fish tank inside the greenhouse is becoming increasingly popular. With a properly sited greenhouse you can grow catfish or tilapia as a summer crop plus trout as a winter crop in most parts of the country, and the vegetables and flowers are a bonus.) The geodesic dome is a popular greenhouse design, sturdy and economical of space.

If you already have a pond, cage culture may be a good approach. A cage size that has been tried and found very practical is four feet in each direction. The framework of the cage is built of lightweight lumber such as 2x2 or 1x2, and covered with Dupont Vexar netting, a nylon material made especially for fish cages. In use, Vexar has been found more durable than metal wire, even the plastic-coated type. You'll need a mesh size small enough to keep your fingerlings from escaping through it but large enough to permit free circulation of water. Quarter-inch mesh will hold any fingerlings; half-inch mesh permits better circulation and should be used if the fish are big enough not to pass through it.

The cage is buoyed by styrofoam floats to keep its topmost twelve inches above the water. This makes access easy for feeding or inspecting the fish. It should have a hinged, screened top to keep the fish from leaping to freedom and to keep fish-eating birds and other predators out.

The cage should be securely moored to keep it from drifting away, and it should have some sort of arrangement to let the owner pull it to shore without having to wade out. Some culturists use an endless loop of rope with a laundry pulley fastened to a stake in the pond and another on a stake driven into the shore.

Fish raised in a floating cage can be stocked at the same density as fish in a tank, and they should be fed by the same rules. Natural food organisms present in the pond may supply additional nutrition, but not enough to justify skimping on the fish chow.

Harvesting could not be simpler: Just pull in the cage, and enjoy your catch.

To be honest, you probably will not save money by raising your own fish intensively. The cost of feed and equipment usually makes the cost per fish about equal to what you'd pay at the supermarket.

Rodale researchers Steve Van Gorder (right) and Jim Fritch harvest channel catfish in the experimental basement pool. With adequate aeration and water quality the fish thrive under these crowded conditions. Water level has been lowered for harvesting the fish; it is normally 18 inches deep.

COURTESY OF RODALE RESOURCES, INC.

However, you'll know that your fish are fresh and free of contaminants (providing you take proper care to keep bug sprays, et cetera, out of their water). And you'll have the satisfaction, by no means negligible, of knowing that you've put the fish on your table by your own work and skill.

More for the Do-It-Yourselfer: Where To Get It

The following partial listing of hatcheries and suppliers of aquacultural equipment has been drawn from the 1980 Buyer's Guide published by *Aquaculture Magazine*. For a complete listing, the guide can be ordered from *Aquaculture Magazine*, P.O. Box 2451, Little Rock, Arkansas 72203. Your local fish and game officer should also be able to assist you in locating suppliers.

Inclusion in this list implies no endorsement by the author, who has attempted to give the reader a broad geographic spread.

Channel Catfish Fingerlings
Schroeder's Fish Farm, Box 598, Carlisle, Arkansas 72024
Fish Breeders, P.O. Box 448, Niland, California 92257
Fish Breeders of Idaho, Route 3, Box 193, Buhl, Idaho 83316
Leon Hill's Catfish Hatchery and Farm, 605 Park Street, Lonoke, Arkansas 72086
Opel's Fish Hatchery, R.R. 1, Worden, Illinois 62097
Smith's Catfish Farms, Gould, Arkansas 71643
Dub Roland Fish Farm, Route 2, Whitesboro, Texas 76273
Walnut Grove Fish Farm, P.O. Box 248, Ripley, Tennessee 38063
Seminole Tribe of Florida, Inc. Aquaculture Project, Route 6, Box 588, Okeechobee, Florida 33472

Elk-Grove-Florin Catfish Farm, 8047 Elk-Grove-Florin Road, Sacramento, California 95823

Eden Fisheries, Inc., Route 2, Box 79, Yazoo City, Mississippi 39194

H-D Fish Farm, Route 1, Cheney, Kansas 67025

Tilapia Fingerlings

Natural Systems, Mike Sipe, President, Route 1, Box 319, Palmetto, Florida 33561

Fish Breeders of Idaho, Route 3, Box 193, Buhl, Idaho 83316

Cypress Bend Fish Ranch, Box 828, Sabinal, Texas 78881

Weisbart & Weisbard, Inc., Fish Department, Route 2, Box 301, Alamosa, Colorado 81101

Rainbow Trout Fingerlings

Blue Springs Trout Farm, Route C, Box 20, Yellville, Arkansas 72687

Brown's Trout Hatchery, Route 362, Bliss, New York 14024 (also brook and brown trout)

Cedar Springs Trout Hatchery, Route 2, Mill Hall, Pennsylvania 17751 (also brook and brown trout)

Beautiful Valley Trout Farm, Inc., Star Route, North New Portland, Maine 04961 (also brook trout)

Crystal Springs Trout Farm, Cassville, Missouri 65625

Fernwood Trout Hatchery, Gansevoort, New York 12831 (also brook trout)

Green-Walk Trout Hatchery, Inc., 36 North 5th St., Bangor, Pennsylvania 18013 (also brook and brown trout)

Jan L. Michalek, P.O. Box 408, St. Louisville, Ohio 43071 (also brook and brown trout)

Musky Trout Hatchery, Inc. Bloomsbury, New Jersey 08804 (also brook, brown, and golden rainbow trout)

Rainbow Acres Trout Farm and Research Station, P.O. Box 29, McIntosh, New Mexico 87032

Trout Haven Ranch, Box 63, Buffalo Gap, South Dakota 57722

Seven Pines Fishery, Lewis, Wisconsin 54851 (also brown trout)

Shenandoah Fisheries, Inc., P.O. Box 276, Lacey Spring, Virginia 22833

Troutdale Ranch, Inc., Gravois Mills, Missouri 65037
Linwood Acres, P.R. 1, Campbellcroft, Ontario, Canada LOA 1BO

Artemia (brine shrimp) egg cysts
Artemia, Inc., P.O. Box 2891, Castro Valley, California 94546
New Technology, U.S.A. Ltd., P.O. Box 524, Bayonne, New Jersey 07002

Macrobrachium Rosenbergii (post-larvae)
Aquafarms, 8167 La Jolla Shores Drive, La Jolla, California 92037
Florida Aquaculture, Inc., Route 1, Box 433X, Arcadia, Florida 33821
Pacific Aquaculture Corp., DEA Fish Farms Hawaii, P.O. Box L Laie, Hawaii 96762
Weyerhaeuser Company, P.O. Box 1584, Homestead, Florida 33030

Aerators
Air-o-Lator Corp., 8100 Paseo, Kansas City, Missouri 64131
Aquatic Eco-Systems, Inc., 1141 North Reams St., Longwood, Florida 32750
Red Ewald, Inc. P.O. Box 519, Karnes City, Texas 78118
Delta Net & Twine Co., Box 356, 619 E. Clay St., Greenville, Mississippi 38701
Hagan Western Fisheries, Box 328, Ft. Collins, Colorado 80522
McCrary's Farm Supply, 114 Park St., Lonoke, Arkansas 72086
Kembro, Inc., P.O. Box 305, Mequon, Wisconsin 53092
Ramco Mat Division, 403 West 21st St., P.O. Box 550, San Pedro, California 90733
Rodale Resources/Otterbine, 1120 South Broadway, Greenville, Miss. 38701, and 576 North St., Emmaus, Pennsylvania 18049

Filtering Equipment
Aquacience Research Corp., 412 East 12th Ave., North Kansas City, Missouri 64116
Red Ewald, Inc. P.O. Box 519, Karnes City, Texas 78118
New Technology, U.S.A., Ltd. P.O. Box 524, Bayonne, New Jersey 07002

James Reed & Associates, 813 Forrest Drive, Box 2159, Newport News, Virginia 23606

Fish Feed

Bucksnort Trout Ranch, Route 1, Box 156, McEwen Tennessee 37101 (catfish and trout)

Sterling H. Nelson & Sons, Inc., Murray Elevators Division, 118 West 4800 South, Murray, Utah 84107 (trout and salmon)

Ralston-Purina Co., Checkerboard Square, St. Louis, Missouri 63188 (catfish, trout)

Allied Mills, Inc., 105 Riverside Plaza, Chicago, Illinois 60606 (catfish)

Rangen, Inc., Buhl, Idaho, 83316 (trout and salmon)

Zeigler Brothers, Inc., Box 95, Gardners, Pennsylvania 17324 (trout and salmon)

Netting for Cages

Conwed Corporation, 332 Minnesota Street, P.O. Box 43237, St. Paul, Minnesota 55164

Inqua Corporation, Box 86, Dobbs Ferry, New York 10522

McCrary's Farm Supply, 114 Park St., Lonoke, Arkansas 72086

Marcrafts, Inc., Flying Point, Freeport, Maine 04032

Nylon Net Company, 7 Vance Avenue, P.O. Box 592, Memphis, Tennessee 38101

Pumps

Aquaculture Research/Environmental Associates, P.O. Box 1303, Homestead, Florida 33030

Aquafarms Canada, Ltd., Feversham, Ontario, Canada

Conde Milking Machine Co., Inc., Sherrill, New York 13461

McCrary's Farm Supply, 114 Park St., Lonoke, Arkansas 72086

Tanks

Aquafarms, 8167 La Jolla Shores Drive, La Jolla, California 92037

Red Ewald, Inc., P.O. Box 519, Karnes City, Texas 78118

Maritime Environment, Inc. 7 Grassy Plain, Bethel, Connecticut 06801

Test Kits (water quality)

Crescent Research Chemicals, 5301 North 37th Place, Paradise Valley, Arizona 85253 (DO, pH, hardness)

Delta Western, P.O. Box 878, Indianola, Mississippi 38751 (DO, pH, hardness)

Hach Chemical Co., P.O. Box 907, Ames, Iowa 50010 (DO, pH, hardness, ammonia)

Hills Kordon, 2242 Davis Court, Hayward, California 94545 (pH)

New Technology, U.S.A., Ltd., P.O. Box 524, Bayonne, New Jersey 07002

Hints on Raising Earthworms for the Backyard Aquaculturist

Since you are not in the earthworm business, a small-scale operation will be sufficient. Twenty-four square feet of surface area should supply all the worms your fish need. While worms can be raised in a specially prepared bed dug in the ground, a container is more convenient and easier to work with. Some authorities recommend a 3-by-8 foot box, but smaller containers that give the same total area will do just as well. The container may be wood or metal; plastic containers are not recommended because many plastics contain toxic substances that may leach into the worm bed and kill the worms.

The bedding should be 12 to 16 inches deep, although you may be able to get away with less. Earthworms need bedding that is porous, resists packing down, and retains moisture. Various materials that professional worm growers favor include shredded newspaper, shredded cardboard, straw, compost, and manure. Peat moss is often mixed in with the bedding to make it more porous.

The bedding should be kept damp but not wet. Check it every day for moisture. Experience will soon show you how much water to sprinkle over it. Turn over the top three or four inches with a pitchfork every two weeks. A garden spading fork will do, but its broad tines may injure some worms.

Earthworms can tolerate a wide range of temperatures above freezing, but they do best between 60° and 80° F. Use a cheap outdoor thermometer to check bedding temperature. Keep the bed out of direct sunlight, which can dry the bedding out surprisingly fast or kill the worms by overheating it.

Commercial worm growers change the bedding every six months. The backyard fish farmer probably will not want the chore of caring for his earthworms over the winter. He will just discard the used bedding and unused worms in the fall after harvesting his fish, and start again in spring. Incidentally, earthworm castings (excrement) are an excellent garden fertilizer.

Worms are sensitive to acidity and alkalinity. The bedding should be kept between pH 6.8 (slightly acid) and ph 7.2 (slightly alkaline). Test the pH every few days with a soil-testing kit. Acidity is easily corrected by sprinkling ground limestone over the bedding; alkalinity is treated by mixing shredded newspaper or *dry* peat moss into the bedding. Because the slowly decomposing bedding materials produce an acid reaction, acidity is usually the problem, and most worm growers routinely sprinkle the beds with ground limestone once a week.

Earthworms sometimes crawl out of their beds *en masse*. An easy way to prevent this is to keep a bright light burning over the bed. Earthworms instinctively avoid bright light and head back down into the bedding.

Feed your worms with vegetable scraps, grass clippings, vegetable trimmings from the grocery store, and dead leaves. Many growers sprinkle a little cornmeal or other fine-ground, starchy food on top of the bedding once or twice a week. Manure is an excellent food, but it must be well-rotted. Fresh manure will ferment, and it may generate enough heat to kill the worms. The manure should also be leached beforehand to wash out salt and animal urine.

The worm bed should be prepared and kept moist for a couple of weeks before you stock it. It can be stocked with earthworms from the local soil or from a manure pile. In 30 to 60 days they should be reproducing—sooner if you can find mature worms. These are recognizable by the broad, swollen band just behind their heads. However, you will probably save time by ordering worms from an established breeder. Two species of earthworms are raised commercially: red worms (*Lumbricus rubellus*) and manure worms (*Helodrilus foetidus*.) Both require a great deal of manure or other organic matter to thrive.

Harvest the worms by turning over the top five or six inches of the bedding with a pitchfork. Disturb only as much of the bedding as you need to—digging up their habitat makes the worms burrow down to the bottom, where they will stay for several days. Save some of the larger worms each time for breeding. Take no more at any time than the fish will eat (based on experience with trial feedings).

Bibliography and Suggestions for Further Reading

Much of the information in this books was drawn from technical papers and personal interviews with professional aquaculturists and fisheries scientists. For general sources, I would suggest the following:

Books

Bardach, John E., Ryther, John H., and McLarney, William O. *Aquaculture: The Farming and Husbandry of Freshwater and Marine Organisms*. New York: Wiley-Interscience, 1972. Probably the best all-round treatment of the subject, although somewhat on the technical side.

Hickling, C. F. *The Farming of Fish*. Oxford, England: Pergamon Press, 1968. Although somewhat out of date with regard to details, this book gives an excellent, simple presentation of basic principles. The author is a retired British fisheries scientist.

Huet, Marcel. *The Textbook of Fish Culture*. West Byfleet, Surrey, England: Fishing News Books Ltd., 1972. As the title states, this is a how-to-do-it book for the professional aquaculturist. The author is a renowned European fisheries expert.

Logsdon, Gene. *Getting Food from Water*. Emmaus, Pennsylvania: Rodale Press, 1978. A breezily written, easy-reading layman's guide to aquaculture. Includes chapters on construction and management of fish ponds as well as less ambitious projects. Also makes discursions into gathering wild foods from waters and wetlands. The author is a regular contributor to several Rodale publications.

Stickney, Robert R. *Principles of Warmwater Aquaculture.* New York: John Wiley & Sons, 1979. A professional book, packed with thorough, detailed, technical information. For those who wish to explore the subject in greater depth and have some scientific background.

Todd, Nancy Jack, *The Book of the New Alchemists.* New York: E. P. Dutton, 1977. In addition to the New Alchemy researchers' original work on fish culture, this book contains informative material on organic gardening, nonchemical pest control, low-energy housing, and utilization of sun and wind energy. It also contains a good deal of philosophizing with which readers may not agree. For more recent results of the New Alchemists' continuous and many-faceted experimentation, the *Journal of the New Alchemists,* published yearly, can be purchased by writing to The New Alchemy Institute, Inc., P.O. Box 432, Woods Hole, Massachusetts 02543.

Periodicals

Aquaculture Magazine (formerly *The Commercial Fish Farmer and Aquaculture News*). Subscription Department, P.O. Box 2451, Little Rock, Arkansas 72203.

Farm Pond Harvest. Professional Sportsmen's Publishing Co., Rural Route 2, Momence, Illinois 60954.

Organic Gardening. Rodale Press, Inc. Emmaus, Pennsylvania 18049.

Salmonid. Published by the United States Trout Farmers Association. P.O. Box 171, Lake Ozark, Missouri 65049.

Index